The Politics of Heredity

SUNY Series,
Philosophy and Biology
David Edward Shaner, editor

THE POLITICS OF HEREDITY

Essays on Eugenics, Biomedicine, and the Nature-Nurture Debate

Diane B. Paul

STATE UNIVERSITY OF NEW YORK PRESS

Production by Ruth Fisher
Marketing by Nancy Farrell

Published by
State University of New York Press, Albany

For information, address the State University of New York Press,
State University Plaza, Albany, NY 12246

Library of Congress Cataloging-in-Publication Data

Paul, Diane B., 1946–
    The politics of heredity : essays on eugenics, biomedicine, and
the nature-nurture debate / Diane B. Paul.
        p.   cm. — (SUNY series, philosophy and biology)
    Includes bibliographical references and index.
    ISBN 0-7914-3821-X (hardcover : alk. paper). — ISBN 0-7914-3822-8
(pbk. : alk. paper)
    1. Human genetics—Political aspects—History. 2. Eugenics-
-Political aspects—History. 3. Nature and nurture—Political
aspects—History. I. Title. II. Series: SUNY series in philosophy
and biology.
    QH431.P34   1998
    174'.25—dc21   97-45212
                                                                CIP

10   9   8   7   6   5   4   3   2   1

CONTENTS

Acknowledgments

My views on the nature-nurture debate have been profoundly shaped and sharpened through a process of friendly (and he might add, "dialectical") struggle with Richard Lewontin. Although our perspectives are sometimes quite different, he has been an unfailingly supportive as well as astute critic of my work. My debt to him is enormous. David Shaner, the series editor, has shown the patience of a saint during the more than four years that it took me to get what might seem a simple job done. Among the many colleagues and friends who, over the years, gave generously of their time and shared their ideas and sometimes unpublished manuscripts, I want especially to thank Garland Allen, John Beatty, Raphael Falk, Jonathan Harwood, Evelyn Fox Keller, Daniel J. Kevles, Philip Kitcher, Robert Proctor, Nils Roll-Hansen, Hamish Spencer, Paul Weindling, and Leila Zenderland. A special debt is owed Michael Ruse, who first urged me to collect these essays in one volume.

A version of chapter 2, "Eugenics and the Left," first appeared in the October/December 1984 issue of *Journal of the History of Ideas*.

Chapter 3, "The Nine Lives of Discredited Data," first appeared in the May/June 1987 issue of *The Sciences* and is reprinted by permission. Individual subscriptions are $21 per year in the U.S. Write to: *The Sciences*, 2 East 63rd Street, New York, NY 10021. A new epilogue incorporates material from "On the Trail of Little

Albert," co-authored with Arthur Blumenthal, which appeared in the Fall 1989 issue of the *Psychological Record*.

The entirety of chapter 4, "The Rockefeller Foundation and the Origins of Behavior Genetics," originally appeared in Keith Benson, Jane Maienschein, and Ronald Rainger, eds., *The American Expansion of Biology* (New Brunswick, NJ: Rutgers University Press, 1991).

A version of chapter 6, "Eugenic Anxieties, Social Realities, and Political Choices," first appeared in the Fall 1992 issue of *Social Research*. This version also incorporates a small amount of material from "Is Human Genetics Disguised Eugenics?," Robert Weir, Susan C. Lawrence, and Evan Fales, eds., *Genes and Human Self-Knowledge: Historical and Philosophical Reflections on Modern Genetics* (Iowa City: University of Iowa Press, 1994).

Chapter 7, "Did Eugenics Rest on an Elementary Mistake?" is an expanded version of "The Hidden Science of Eugenics" (co-authored by Hamish Spencer), which appeared in the March 23, 1995 issue of *Nature*. Chapter 7 is also to be published under the same title by Cambridge University Press in *Thinking about Evolution: Historical, Philosophical, and Political Perspectives*, Rama Singh, Costas Krimbas, Diane Paul, and John Beatty, eds.

The entirety of chapter 8, "From Eugenics to Medical Genetics," first appeared in the Winter 1997 issue of the *Journal of Policy History* (Volume 9, No. 1), 96–116. Copyright 1997. Reproduced by permission of the publisher.

A version of chapter 10, "PKU Screening: Competing Agendas, Converging Stories," originally appeared in Michael Fortun and Everett Mendelsohn, eds., *The Practices of Human Genetics* (Dordrecht: Kluwer Academic Publishers, 1998).

1

INTRODUCTION

I

By the end of the Spring semester in 1975, I badly needed a break. For four years, I had been teaching three or four courses a semester at the University of Massachusetts at Boston while writing my dissertation, "The Politics of the Property Tax," and readying it for publication. So, to reward myself, I signed up to work and study at the UMass field station on Nantucket Island. I agreed to mend nets, wash bottles, collect algae, cook, and do a host of other odd jobs in return for the opportunity to indulge an old enthusiasm for natural history.

Linking my political science and natural history interests was the last thing on my mind. On the contrary, the appeal of a summer's work in field biology lay in its very distance from my academic interests, then focused on political economy, public policy, and the history of political theory.

By the time the summer ended, my enthusiasm for natural history had turned into an interest in biology. On returning to the university, I registered for a formal course in the subject. And then for another. And another. Each time I told myself that I would just take one more course, but I couldn't stop. Eventually it became

1

clear that I had to find a way to make these apparently disparate interests mesh: I told my astonished Department that I intended to return to school and, afterward, to write about science rather than taxes. That they tenured me anyway was an act of faith that I hope the essays in this volume go at least some way to justify.

In the years that followed, my disciplinary identity shifted from political scientist to historian of science. But my interests in public policy and political theory did not fade. They continue to inform, and I hope enrich, all my writing on the history of evolution and genetics.

The essays collected in this volume (as well as others written over the same period) explore specifically political dimensions of science. Of course, much recent work in science studies is also concerned with the political. Numerous ethnographic studies of laboratory life and detailed historical case studies have focused on the ways in which truth may be politically "negotiated," often emphasizing the role of social interactions in determining scientific success. Steven Shapin has recently summed up the body of work in the sociology of scientific knowledge and related history and philosophy "as concerned to show, in concrete detail, the ways in which the making, maintaining, and modification of scientific knowledge is a local and mundane affair."[1] But contrary to the old adage, not all politics is local.

Marxists have always recognized that fact. However, many scholars are made understandably nervous by accounts of scientific change and the reception of scientific ideas that invoke large sociopolitical forces. They associate such accounts with generalizations about society and class that are too broad to be defensible. But as I hope these essays help to demonstrate, analyses of politics outside the laboratory need not be crude. My aim is to contribute to a growing body of historical work on scientific knowledge and practice that is simultaneously broad in scope, politically relevant, and sensitive to nuance.

II

The essays collected in this volume, which were written over a period of a dozen years, explore the history of eugenics, biomedicine, and the nature-nurture debate. Many show how political factors underlie various apparent (and otherwise inexplicable) changes in scientific and biomedical perspectives. They address such questions as the following: Why did assumptions about the role of genes in human behavior that were taken for granted in the 1960s come under vehement attack in the 1970s? Why did the same scientific principles that in the 1930s seemed to demonstrate a desperate need for eugenics come in the 1950s instead to explain its futility? Why was the distinction between good and bad eugenics abandoned in favor of a distinction between medical genetics and the "pseudo science" of eugenics? Why was carrier detection viewed in the interwar period as a means to root out defective genes and in the postwar period as a means to mask their effects? How was the history of screening for phenylketonuria transformed from the tale of a troubled program to the greatest success story of applied human genetics? But the essays also show that some apparently profound shifts were quite superficial; that changes in rhetoric may obscure the stability of core underlying beliefs.

The first essay in the collection, "Eugenics and the Left," was published in 1984. It contests the then-conventional association of socialism with opposition to eugenics, noting that Marxist and Fabian condemnation of the race and class bias of the mainstream movement should not be equated with in-principle opposition to the rational control of reproduction. Indeed, to many Left intellectuals, especially in the sciences, eugenics seemed to follow logically from the rejection of laissez-faire. In the 1930s, 1940s, and 1950s, geneticists of every political persuasion were convinced that individuals varied in their genetic value and that the worthiest should be encouraged to have more children and the least worthy fewer or none.

A second argument challenges the (still common) claim that the post–World War II demise of eugenics is explained by the progress of science. It argues that the scientific findings said to have undermined eugenics in fact occurred much too early to account for postwar developments. In a recent essay, co-written with Hamish Spencer, entitled "Did Eugenics Rest on an Elementary Mistake?" I examine that argument with respect to the Hardy-Weinberg theorem (see chapter 7 in the present volume). An implication of that theorem is that, when genes are both recessive and rare, the number of carriers will vastly exceed those actually affected. Since segregation and sterilization do not reach the clinically asymptomatic, it might seem that programs of eugenical selection are beside the point. But contrary to the conventional wisdom, few geneticists drew this conclusion. Instead, they saw the lesson of Hardy-Weinberg to be the need for research to detect apparently healthy carriers and prevent them from breeding. Nor is there any scientific reason that this theorem should have carried the same implications for them as it does for us. The lessons that now seem so plain follow only in the context of values that, though widely held today, were disdained by an earlier generation of geneticists.

Shifts in the meaning of "eugenics" and the struggles to demarcate it from genetics are explored in a number of essays. In "The Rockefeller Foundation and the Origins of Behavior Genetics," we see that genetics has sometimes been equated with basic research and eugenics with applied research. Thus, to Foundation officers even a proposed Institute for Race Biology, the ultimate aim of which was to improve the "biological constitution" of the population, had nothing to do with eugenics. "Eugenic Anxieties, Social Realities, and Political Choices" identifies other conventional lines of demarcation, including motivation (where eugenics is equated with social goals, whereas medical genetics is identified with individual aims) and means (where eugenics is equated with coercion, whereas medical genetics is associated with freedom of choice). Reflecting my interests in political theory, this essay also analyzes the protean meanings of freedom and coercion, explores how "autonomy" came to trump every other value in the sphere of biomedicine, and probes some of the consequences.

"Eugenic Origins of Medical Genetics" provides a historical perspective on efforts to demarcate eugenics from other practices and also challenges the conventional view that eugenics fell into disrepute following World War II. In the 1950s and 1960s, medical geneticists often characterized their work as "eugenics"—though of a kind sharply distinguished from the "bad" eugenics of the past. But by the 1970s, the term had become disreputable. As a consequence, geneticists largely abandoned their attempt to distinguish good eugenics from bad. Medical genetics was now contrasted instead with the "pseudo science" of eugenics. At the same time, critics moved in the opposite direction, toward expansive definitions that associate genetic medicine with a host of now-despised practices.

"Genes and Contagious Disease" extends the arguments both of "Eugenic Origins of Medical Genetics" and "Did Eugenics Rest on an Elementary Mistake?" It explores what happened when eugenicists' hopes of identifying clinically asymptomatic carriers were finally realized. In the 1950s, the first reliable methods of carrier detection were employed not to efficiently root out defective genes, but rather to mask their effects—a strategy to reduce the immediate burden of genetic disease at the cost of increasing the incidence of disease-causing genes. The essay traces the scientific and social developments that explain why the new technology was turned in such a different direction than eugenicists had once expected. It also illustrates the difficulty of agreeing on a definition of eugenics. From one perspective, to embrace a masking strategy is to abandon eugenics, defined in this case as a concern with the future of the gene pool. From another perspective, it is seen to mark a turn from one kind of eugenics to another.

The final essay in this collection, "PKU Screening: Competing Agendas, Converging Stories," concerns the first treatable genetic disease: phenylketonuria, or more simply, PKU. In 1948, the British geneticist Lionel Penrose used the disease to illustrate the futility of eugenical selection against rare genes. PKU has ever since been employed as a potent symbol.

Since the early 1960s, it has been possible to identify affected newborns, who can be placed on a diet that prevents the severe mental retardation associated with the disease. This therapeutic success is frequently cited by enthusiasts for genetic medicine, for whom it illustrates the good that screening can do. But the same case is as often invoked by skeptics of genetic medicine. For them, the treatability of the disease carries a different message—that genes are not destiny. While they draw disparate lessons from the PKU case, enthusiasts and critics have a joint interest in portraying treatment as simple and wholly effective. Alas, it is neither. As a result of these converging interests, an appealing but deceptive story about PKU has become entrenched in popular literature. And as "The Nine Lives of Discredited Data" shows, once entrenched, stories that carry moral or political messages are almost impossible to uproot.

III

These essays are political in more than one sense of the term. Most obviously, they are *about* the connections, at various levels, between politics and science. But they are also intended, in the broadest sense, as political interventions. I have entertained the hope that such historical work might be relevant to contemporary debates about public policy. This political commitment has entailed efforts to write for disparate audiences: the general public as well as specialists, scientists as well as historians, and individuals who would contest, as well as those who share, my own broadly socialist perspective. To this end, I have written for journals as diverse as the *Journal of the History of Ideas, Science for the People, Newsweek, Scientific American,* the *Quarterly Review of Biology,* and *Nature.* Indeed, this diversity is one reason for collecting some of the essays in one volume.

For some time now, I have been distressed by the increasingly self-referential turn that much academic work in science studies seems to be taking. The influence of postmodernism has been productive in some respects but counterproductive in others. It has taught us the important lesson that the truth of our views is guaranteed neither by God nor by Nature nor by the laws of history. While the recent *Social Text* affair has generated a certain amount of romantic nostalgia for the old-fashioned, plain-speaking academic Left,[2] postmodernist excesses cannot be countered by reasserting old certainties.[3] There are good reasons why foundationalism has fallen from favor. But taken to extremes, the romance with "theory" has rendered much work in science studies unintelligible both to scientists and the public.

At the same time, many scientists confuse the purpose of historical and sociological studies with advocacy: they expect deference from outsiders and dismiss even the most plain-speaking scholars who fail to oblige.[4] It is not easy to speak across this science/humanities divide. Moreover, the mutual incomprehension that characterizes the current "science wars" has its analogue in the politics of many nonscientific debates. On the one hand are those who believe that they possess a privileged access to truth. On the other are those who have no grounding at all for their opinions; their claims are reduced to expressions of taste. Either stance renders political discussion pointless. There is no arguing either with ordained Truth or with likes and dislikes. We do not judge subjective preferences to be right or wrong, valid or invalid.[5] But these essays were written on the assumption that there remains a role for reasoned argument in current debate, that it is still worth trying to reach across the profound and, unfortunately, deepening professional and political divides.

IV

Given my background and professional and political aims, I have naturally been interested in a related issue that has been the subject of passionate debate within the science studies community: Is a wholly social epistemology compatible with moral and political critique? It has often been charged that the symmetry principle—which holds that true and false beliefs need equally to be explained sociologically—undermines or even precludes an evaluative stance. Is the accusation justified?

The claims of such sociology of knowledge (SSK) practitioners as David Bloor, who adopts a stance of "moral neutrality,"[6] or Harry Collins, who asserts that SSK leaves science exactly as it is,[7] certainly lend it credence. In *The Golem: What Everyone Should Know about Science,* Harry Collins and Trevor Pinch tell the story of cold fusion. Their tale revolves around scientists and administrators whose motivations are crassly commercial. The desire to patent and market the discovery leads scientists to hype their results, which are announced at press conferences rather than in journals, and to withhold details of their experiments. But Collins and Pinch do not draw the conventional moral. If the market has fully penetrated the scientific arena, that is fine with them. "It is our image of science which needs changing," they conclude, "not the way science is conducted."[8]

Such an apolitical stance is unappealing to many SSK scholars. As both partisans and critics of current trends in science studies have noted, a "self-conscious radicalness" informs much of the literature, indicating that its authors are motivated at least in part by normative considerations. Bruno Latour writes of many postmodern critics of science that "they maintain the will to denounce and debunk, but have no longer any grounds to do so."[9] Politically engaged practitioners, on the other hand, insist that methodological relativism does not require evaluative restraint; indeed, that SSK can and should contribute to progressive or "emancipatory" goals. They believe these aims to be best served by a focus, not on whether scientific views are true or false, but on whether they are empowering or disempowering. In their view, it is possible to judge theories as emancipatory or oppressive, independent of any evaluations about their fit with the world.[10] They want "not truth, but justice."[11]

That approach in effect reduces science, not only to politics, but to a highly subjective sort of politics. It thus ensures that critique will be solipsistic. Judgments about science would have meaning that is internal only to groups whose members already agree. For political minorities to be effective, they must show that those they criticize have violated widely shared norms. In arguing that Cyril Burt's work was sloppy or faked—that it was bad science—Marxists were able to appeal to individuals with very disparate political convictions. Does anyone think that labelling Burt's work "disempowering" would have been equally effective? Preaching to the already converted sometimes has a point. But for minorities, simply rallying the troops will never win wars. Jane Flax argues that "there may be more effective ways to attain agreement or produce change than to argue about truth. Political action and change require and call upon many human capacities including empathy, anger, and disgust."[12] But the efficacy of appeals to emotion cannot and should not be dissociated from beliefs about what is, in fact, the case. Even if we were comfortable with the view that scholarship should simply subserve politics, how will minorities provoke these politically useful emotions if they are barred from grounding appeals for justice in "reason, knowledge, or truth"?

Reducing science to politics is not a new endeavor. In the 1940s and 1950s, orthodox Marxists argued that the natural sciences are superstructural in the same way as politics and law. It seemed to follow that scientific theories serve the interests of either proletarians or the bourgeoisie. Looking back on the role of French communist intellectuals in the early days of the Cold War, Louis Althusser wrote:

> In our philosophical memory it remains the period of intellectuals in arms, hunting out error from all its hiding-places; of the philosophers we were, without writings of our own, but making politics out of all writing, and slicing up the world with a single blade, arts, literature, philosophies, science with the pitiless demarcation of class—the period summed up in caricature by a single phrase, a banner flapping in the void: "bourgeois science, proletarian science."[13]

The theory of two sciences found its most brutal expression in the condemnation of genetics (and geneticists) as "Menshevizing," "idealist," and "undialectical."

In part reacting to that history, many Marxists have adopted an (Althusserian) approach that wholly dissociates science and politics. In this perspective, good science is uncontaminated by ideology. Associating themselves with the cause of science, Marxists have often denied the political dimensions of their own work. But if they are naive to view politics simply as a contaminant, they are right to resist the claim that science *is* politics. Unlike politics (or religion or art or, to take a recent famous example, baseball[14]) the *raison d'être* of science is the understanding of nature. To omit nature from our evaluations is to render the enterprise unintelligible. But there is a political reason as well to judge theories by their "fit with the world." For the Left, as for any minority, to accept the alternative is to court disaster.

Notes

1. Steven Shapin, "Here and Everywhere: Sociology of Scientific Knowledge," *Annual Review of Sociology* 21 (1995): 304.

2. Katha Pollitt, "Pomolotov Cocktail," *The Nation*, June 10, 1996.

3. Alan Sokal, "A Physicist Experiments with Cultural Studies," *Lingua Franca*, May–June 1996, pp. 62–64.

4. Cf. Peter Gross and Norman Levitt, *Higher Superstition: The Academic Left and Its Quarrels with Science* (Baltimore: Johns Hopkins University Press, 1994); Louis Wolpert, *The Unnatural Nature of Science: Why Science Does Not Make (Common) Sense* (London: Faber and Faber, 1992).

5. Charles Taylor, *Philosophical Arguments* (Cambridge: Harvard University Press, 1995), p. 254.

6. *Knowledge and Social Imagery* (London: Routledge and Kegan Paul, 1976), p. 10.

7. "An Empirical Relativist Programme in the Sociology of Scientific Knowledge," in K. Knorr-Cetina and M. Mulkay, eds., *Science Observed: Perspectives on the Social Study of Science* (London: Sage, 1983), p. 98.

8. Harry M. Collins and Trevor Pinch, *The Golem: What Everyone Should Know about Science* (Cambridge: Cambridge University Press, 1993), p. 78.

9. *We Have Never Been Modern*, trans. Catherine Porter (Cambridge: Harvard University Press, 1992), p. 291.

10. See Jane Flax, "The End of Innocence," in J. Butler and J. W. Scott, eds., *Feminists Theorize the Political* (London: Routledge, 1992); William T. Lynch, "Ideology and the Sociology of Scientific Knowledge,"

Social Studies of Science 24 (1994): 198; Kelly Oliver, "Keller's Gender/ Science System: Is the Philosophy of Science to Science as Science Is to Nature?" *Hypatia* 3 (1989): 137–48; Bruce Robbins and Andrew Ross, "Mystery Science Theater," *Lingua Franca,* July–August 1996, p. 57; Steve Woolgar and Keith Grint, "A Further Decisive Refutation of the Assumption that Political Action Depends on the 'Truth' and a Suggestion that We Need to Go beyond This Level of Debate: A Reply to Rosalind Gill," *Science, Technology, and Human Values* 21 (1996): 353–57.

11. Flax, as cited by Rosalind Gill, "Power, Social Transformation, and the New Determinism: A Comment on Grint and Woolgar," *Science, Technology, and Human Values* 21 (1996): 351.

12. Flax, p. 458.

13. *For Marx* (London: Allen Lane, 1969), p. 1.

14. Stanley Fish, "Professor Sokal's Bad Joke," *New York Times,* May 21, 1996, p. A23.

2

EUGENICS AND THE LEFT

Preface

As my interest in history of science took root, the first questions that captured my attention were those directly related to topics with which I was already engaged as a political scientist—in particular, the history of Marxism. Three early essays (not reprinted in this volume), "Marxism, Darwinism, and the Theory of Two Sciences," " 'In the Interests of Civilization': Marxist Views of Race and Culture in the Nineteenth Century," and "A War on Two Fronts: J. B. S. Haldane and the Response to Lysenkoism in Britain," explored aspects of Marxism's troubled relationship with the natural sciences.[1]

In the course of research for these essays, I found that many Marxist intellectuals, and especially geneticists, would be considered "hereditarians" by the standards then prevailing on the Left. Even more surprising, they were often advocates of some form of eugenics. That eugenics appealed to political and social radicals, as well as conservatives, is now a commonplace. In his ground-breaking book *In the Name of Eugenics,* which compared eugenics movements in Britain and the United States, Daniel Kevles stresses the point,[2] and it has been reinforced since by work on eugenics elsewhere, including Russia, Germany, and especially Scandinavia. But at the time, the discovery came as a shock.

Perhaps it should not have, since both Mark Haller's *Eugenics: Hereditarian Attitudes in American Thought* and Kenneth Ludmerer's *Genetics and American Society* had discussed eugenics' appeal to political progressives and to geneticists respectively. But by the time I turned to the subject, eugenics had been labelled a reactionary "pseudo science." It followed for most scholars writing in the 1970s and 1980s that neither Marxists nor geneticists could be counted among its supporters. Ludmerer had distinguished good eugenics from bad. But he wrote just as the concept of a "good eugenics" was becoming an oxymoron.

The present essay has been edited to eliminate passages that now seem redundant or digressive, correct errors, and clarify a few points. However, I have resisted the temptation to make major substantive changes. Were I to write an essay on the same theme today, I would modify my claim that a consensus among geneticists on the in-principle desirability of eugenics collapsed in the 1950s and 1960s. Work I have done since indicates that during this period many geneticists were actually quite comfortable with the eugenics label. Indeed, excitement generated by the discovery of the genetic code apparently combined with anxieties provoked by atmospheric nuclear weapons tests and increased medical exposures to ionizing radiation to produce a resurgence of interest in eugenics. (This claim is defended in chapter 8, "Eugenic Origins of Medical Genetics.") Not until the 1970s did it become a term of derision even for scientists. (It was also in the 1970s that both scientists and historians came to view eugenics as a "pseudo science").

I would also adopt a different tone toward the socialist geneticists who rejected a laissez-faire approach to reproduction. At the time, I did not treat their views with much respect. As will be evident from later essays (see, for example, "Eugenic Anxieties, Social Realities, and Political Choices"), I would no longer dismiss out of hand their assertion of a social interest in reproductive decisions. Left scientists may have had too little concern for individuals and too much faith in the state, but in my view they were right to insist that individual decisions do have social consequences that are matters of legitimate social concern.

I now see the essay as unintentionally illuminating about the changing character of the nature-nurture debate. In 1984, I labeled as "hereditarian" or "biological determinist" the view that differences in mentality and temperament were substantially influenced by genes—employing these terms as though their meanings were unproblematic. That usage today would surely be contested. For the view implicitly disparaged by these labels is once again widely accepted by scientists and the public alike.

Introduction

"The dogma of human equality is no part of Communism . . . the formula of Communism: 'from each according to his ability, to each according to his needs' would be nonsense, if abilities were equal."[3] So asserted J. B. S. Haldane, the distinguished Marxist geneticist, in the *Daily Worker* of November 14, 1949. Even at the height of the Lysenko controversy—and writing in the newspaper of the British Communist Party (on whose editorial board he served)— Haldane refused to retreat from the positions regarding the existence of innate human inequalities and the value of a socially responsible eugenics with which he had been associated since the 1920s. Indeed, Haldane would maintain these views, in only slightly modulated form, until his death in 1964.[4]

If Haldane's opinions were *sui generis*, they would be of only minor interest. But in fact, his linked beliefs in socialism, inequality, and eugenics were widely shared on the Left, particularly among Marxists and Fabians with scientific interests. Beatrice and Sidney Webb, George Bernard Shaw, Havelock Ellis, Eden and Cedar Paul, H. J. Laski, Graham Wallas, H. G. Wells, Edward Aveling, Julian Huxley, Joseph Needham, C. P. Snow, H. J. Muller, and Paul Kammerer—to note just some of the more prominent figures—all advocated (though in varying forms: some "positive" and some "negative," some here-and-now and some only in the socialist future) the improvement of the genetic stock of the human race through selective breeding.[6] It was Shaw who argued that "there is now no reasonable excuse for refusing to face the fact that nothing but a eugenic religion can save our civilization." Eden Paul asserted that "unless the socialist is a eugenicist as well, the socialist state will speedily perish from racial degradation." And H. J. Laski proposed that "the different rates of fertility in the sound and pathological stocks point to a future swamping of the better by the worse. As a nation, we are faced by race suicide."[7]

The history of eugenics has been presented so often as though it were simply the extension of nineteenth-century social Darwinism[8] that we have nearly lost sight of the fact that important segments of the Left (as well as the women's movement) were once also enthusiastic about the potential uses of eugenics.[9] Indeed, in Britain and the United States there once existed a movement known as "Bolshevik Eugenics." In both countries, the enthusiasm that many biologists, like their colleagues in other disciplines, felt for the Soviet Union was rooted in their conviction that it would spur scientific development and promote a scientific perspective. For the biologists, one test of a scientific outlook was a society's willingness to adopt a genuinely scientific stance toward questions of what used to be called "race betterment." Marxist and Fabian biologists believed that Western societies had largely failed this test. To the extent that eugenic sentiment had taken hold, it was used in a pseudo-scientific way to buttress the conventional social order and provide a scientific gloss on racial and class prejudices. There could be no valid comparison of the intrinsic worth of different individuals, they asserted, in a class-stratified society.

That individuals differed significantly in their genetic endowments, not just in respect to physical characteristics or even intelligence but also in respect to specific traits of character and personality, was taken for granted. It was also assumed that the fitter should be encouraged, and the less fit discouraged, from reproducing; and that such a policy could only be successfully pursued in a society that provided approximately equal opportunities to all its members. That the Soviet Union was perceived as such a society, and hence promised to provide the first socially responsible opportunity to test and apply eugenical principles, helps explain its appeal to scientists.[10]

Were it not for widely held assumptions regarding what Right and Left must stand for, this would not be surprising. After all, Social Darwinism was associated in Britain and the United States with the classical liberal commitment to unrestricted laissez-faire and emphasis on individual choice while eugenics implied, at a minimum, the development of a social, and often a state, concern with reproduction. As Sidney Webb wrote: "No consistent eugenist can be a 'Laisser Faire' individualist unless he throws up the game in despair. He must interfere, interfere, interfere!"[11] Lancelot Hogben similarly remarked that "the belief in the sacred right of every individual to be a parent is a grossly individualistic doctrine surviving from the days when we accepted the right of parents to decide whether their children should be washed or schooled."[12] On the other hand, acceptance of "social consciousness and responsibility in regards to the production of children"[13] and, even more, state action to enforce that responsibility, ran counter to the philosophic temper of classical liberalism as well as to Catholic doctrine.[14] (The Catholic philosopher G. K. Chesterton denounced eugenics as an aspect of the "modern craze for scientific officialism and strict social organization.")[15] In *Heredity and Politics* (1938), Haldane himself insisted that attitudes toward eugenics did not correspond with the usual Left/Right political divisions. "The questions with which I shall deal cut right across the usual political divisions," he wrote. "For example, the English National Council of Labour Women had recently passed a resolution in favour of the sterilization of defectives, and this operation is legal in Denmark and other countries considerably to the 'left' of Britain in their politics."[16] Donald MacKenzie expresses the prevailing view of the history of eugenics and the Left when he writes that "the radical scientists of the 1930s saw the eugenics movement as a paradigm case of the anti-working class use of science, and the defeat of eugenic ideology became one of their major preoccupations"—and cites Haldane and Hogben as examples.[17] But Haldane and Hogben aimed to reform eugenic ideology, not defeat it.

Many Marxists and Fabians were indeed critics of certain kinds of eugenics. I do not mean to minimize the differences between the eugenics of the Right and that of the Left or, for that matter, distinctions among individuals and groups on the Left. Thus Hogben favored the California sterilization laws whereas Haldane opposed them. Haldane and Hogben, as well as Huxley, Jennings, Needham, Muller, and other geneticists active in the interwar period are rightly characterized as critics of the organized eugenics movement of their own day, which they criticized for its simplistic science and social bias. But the virtually exclusive focus on their critical role has served to obscure the fact that these geneticists were egalitarians only in a restricted sense. Haldane was fond of quoting Engels's assertion: "The real content of the proletarian demand for equality is the demand for the abolition of classes. Any demand for equality which goes beyond that, of necessity passes into absurdity."[18] Socialist geneticists, like their colleagues of other political persuasions, assumed that differences in intelligence, personality, and character were strongly influenced by heredity. On the Left, this assumption combined with statist and scientistic leanings to make eugenics seem a matter of common sense.[19]

Socialist scientists and science popularizers saw themselves as engaged in a struggle for the cause of science and materialism against the forces of obscurantism; they shared a common conception of the progressive world as one of electricity and machinery, and they welcomed an enormously expanded role for the state. The "social relations of science" movement of the 1920s and 1930s reflected the assumption of many at the time that the causes of science and socialism were inextricably linked.[20] And socialism, for the British and American scientists associated with this movement, as well as for many non-scientific socialists, not only presupposed but, for some, was essentially constituted by a substantial increase in the authority of the state. Few would perhaps go so far as Karl Pearson, who wrote:

The legislation or measures of police, to be taken against the immoral and antisocial minority, will form the political realization of Socialism. Socialists have to inculcate that spirit which would give offenders against the State short shrift and the nearest lamp-post. Every citizen must learn to say with Louis XIV, *L'état c'est moi!*[21]

But neither were socialists in general reluctant to see the state involve itself in formerly private spheres of life. Marx may have ridiculed those who equated socialism and nationalization, but it was not his works that led so many scientists to socialism in the 1920s and 1930s. Rather it was the example of the Soviet state—its commitments to planning, to technical efficiency, to science education and research—that had such influence. However sharp in theory the distinction between Fabianism and Marxism, in practice it has often been blurred (as in the person of J. D. Bernal, who belonged both to the Fabian Society and the British Communist Party).

The focus of this essay is on one particular group of socialists: Anglo-American scientists who, in the 1920s, 1930s, and early 1940s, aimed to develop a socially responsible eugenics; that is, a program to be implemented in a society that had abolished social classes and hence could truly differentiate between the effects of heredity and environment. It therefore represents another chapter in the stories of both the social relations of science and of the eugenics movements. But contributing to a fuller and more adequate account of those movements is not its only, or even its principal, aim. That is to demonstrate the existence, by the 1920s, of a consensus among geneticists concerning the role of heredity in the determination of intellectual, psychological, and moral traits so complete that virtually no one—including Marxist and other Left geneticists—is to be found outside it. In his otherwise admirable biography of T. H. Morgan, Garland Allen asserts that in the early stages of genetics "eugenicists increasingly claimed that personality traits, intelligence, and behavior patterns were genetically determined—claims most geneticists realized had no basis in fact."[22] It would be closer to the mark to say that this was a claim that almost no one doubted. And from this assumption, it was but a short (if not a logically necessary) step to the advocacy of eugenics.

Most striking is the speed with which this consensus collapsed. Assumptions that appeared self-evident to most geneticists in the mid-1930s found but a handful of defenders two decades later. The ridicule heaped on Robert Graham's proposal to artificially inseminate women with the sperm of Nobel Prize winners is a particularly striking example of the distance travelled. In 1935, similar proposals by the Marxist geneticists H. J. Muller (after whom Graham's sperm bank was to have been named) and Herbert Brewer were widely approved by their peers.[23] Thus Haldane offered Brewer the use of his name, his money, and even his gametes.[24] And such doubts as were expressed by professional colleagues were unrelated to the assumption of substantial genetic determination of intelligence, personality, and character. This response immediately raises another question: why, within the space of perhaps two decades, should a scientific consensus have collapsed? The answer, I will suggest, has little to do with events internal to the science itself. But first let us look in some detail at the content and original reception of Muller's proposal.

H. J. Muller and "Bolshevik Eugenics"

H. J. Huller was the scientist most prominently associated in the 1920s and 1930s with the development of a socialist eugenics. His book, *Out of the Night,* essentially completed in 1925 but first published a decade later and distributed in England by the newly formed Left Book Club, was in effect its manifesto. Muller assumed that intelligence, character, and personality have an irreducible and substantial genetic basis. Without a proper environment, Muller argued, the best genes will be wasted, but even the best environment cannot turn an inherently stupid or selfish person into one who is intelligent or altruistic. The ideal situation is one in which favorable environments allow the expression of superior genotypes. "There can be no doubt," he wrote, "that mankind must be highly variable in regard to genes which determine the original physical basis of emotional and temperamental as well as more purely intellectual traits . . . not only the mere presence or absence of the trait, and its time of appearance, but also its intensity and many details of its mode of expression must be influenced by the genes, just as we find to be true of physical characteristics of the blood, the hair, the teeth, and all other parts of the body. In no way does this contradict the fact . . . that environment also is of the utmost importance in the development of the mental superstructure."[25]

In capitalist societies, he argued, genetic merit cannot be distinguished from environmental good fortune. Only at the extremes of feeblemindedness and genius is it possible to conclude with any certainty that the particular genotype is deficient or superior. This blurring of the effects of heredity and environment, and our consequent inability to locate and make use of superior genotypes, is only avoidable in a society offering equal opportunities to all its members. The bourgeoisie will not voluntarily relinquish its privileges, so a social revolution is needed.[26] After that revolution (and currently in the Soviet Union), enormous opportunities for the genetic improvement of mankind will be available.

But for these to have effect, child-bearing and raising must first be made attractive. Work opportunities outside the home should be opened to mothers, who must be allowed to limit the size of their families through the legalization of birth control information and devices (with abortion as a second line of defense), the pain of childbirth, ignored "because the doctors have been mostly men, who regard such pains in women as obligatory, or even sadistically look upon them as desirable,"[27] could be mitigated and the illnesses of childhood relieved, and public organizations developed to aid in cooking, laundering, and other aspects of child-raising and housework. A radical improvement in women's position will be accompanied by the disappearance of superstitions and taboos regarding

family relationships and sex. As a result, considerations of reproduction will be divorced from those of love.[28]

In this transformed environment, it would be possible to apply
new inventions and discoveries in biology to the control of what
had hitherto been the female's child-bearing role, thus vastly increasing "both the possibilities of eugenics and our ability to order
these processes in the interests of mother and child."[29] These techniques, which include the transplantation and consequent development of the fertilized egg from one female to another and the
development of human eggs without fertilization (that is, without
a father), would "greatly extend the reproductive potencies of females possessing characters particularly excellent, without thereby
necessarily interfering with their personal lives." The ultimate ideal
would be ectogenesis, or the complete development of the egg outside the mother's body (an idea originally advanced in Haldane's
Daedalus and taken up by J. D. Bernal in *The World, the Flesh,
and the Devil,* books much admired by Muller).[30]

But even with present techniques it would be possible to
artificially inseminate many women with the sperm of "some tran-
scendently estimable man"; that is, men superior in intelligence
and "highly developed social feeling—call it fraternal love, or sympathy, or comradeliness, as you prefer."[31] Offspring from such
matings could be expected to stand, on average, halfway in their
heredity between their fathers and the average of the population.
Hence, if it were not for social prejudice and inertia, we could right
now dramatically raise the intellectual and moral level of the population. "It is easy to show," wrote Muller, "that in the course of a
paltry century or two . . . it would be possible for the majority of the
population to become of the innate quality of such men as Lenin,
Newton, Leonardo, Pasteur, Beethoven, Omar Khayam, Pushkin,
Sun Yat Sen, Marx (I purposely mention men of different fields and
races), or even to possess their varied faculties combined."[32]

Although we presently have the technical ability to effect this
change, it would almost certainly be abused in capitalist society.
Directed by the same forces that control our press and public opinion, the new system would produce not men of the highest type but
rather the "maximum number of Billy Sundays, Valentinos, Jack
Dempseys, Babe Ruths, even Al Capones. . . . "[33] Fortunately, our
society is at present anyway disintegrating. In the absence of an
economic and social revolution, there will be no revolution in our

attitudes toward sex and reproduction; with one, we will naturally aim to produce Lenins and Newtons.[34] As we will see, while some doubted that what the world needed was more Lenins (or even Newtons), few thought that there was any substantial scientific barrier to that project.

The Reception of Muller's Book in the Soviet Union and the West

Muller's twin enthusiasms for socialism and eugenics prompted his emigration to the Soviet Union in 1934. There he worked with N. I. Vavilov at the Institute of Genetics in Moscow and completed *Out of the Night,* which he presented, along with an effusive letter of personal appeal, to Stalin. Not only his choice of recipients but his timing could not have been worse since genetics had already come under attack as inextricably linked to eugenics, and eugenics to the old social order. There was some real basis to this charge. As Loren Graham has argued (and Theodosius Dobzhansky noted somewhat reluctantly in his memoirs and in his private correspondence with Muller), the eugenics movement in the Soviet Union had a decided class bias, at least up to 1925. In the early 1920s, the vast majority of scientists were drawn from that class generally hostile to the October (if not to the February) Revolution. Their overriding concern was with what they saw as the dysgenic effects of the Revolution, civil war, and especially emigration, which had decimated the aristocratic and intellectual classes. They collected and published genealogies of aristocratic families and those of high achievement and issued dire warnings about the consequences of continued biological degeneration.[35]

It is probable that geneticists would have come under attack during the "cultural revolution" of the 1920s even in the absence of any link to eugenics. The 1920s witnessed an upsurge of "Lamarckian" sentiment in the Soviet Union that was directed against plant and animal as well as human genetics and whose roots were found in the revolutionary optimism of a public (particularly radical students in the universities) to whom everything seemed possible to those with the will to make it so. It is hardly surprising that the newly revolutionized students and workers were more attracted to "Lamarckism" than to the views associated with Mendel, Weismann, and T. H. Morgan. But genetics' association with a eugenics movement nostalgic for the old social order was an added burden that contributed to its collapse.

As public, and also official, sentiment turned increasingly against them, the Russian eugenicists either turned to non-human genetics or to a reformulated eugenics, in which genealogies of outstanding proletarians replaced those of aristocrats. This feint fooled virtually no one, excepting the eugenicists themselves. That was the situation when Muller arrived in the Soviet Union. It was not long before he realized that conditions for the development of a Bolshevik eugenics were less promising than he had assumed. But he decided that negative sentiment could be overcome by restating the eugenical case in more tactful terms and by going directly to the top; i.e., to Stalin, whom he (mistakenly) believed to be sympathetic to his aims.[36] Hence, *Out of the Night* was written to flatter a Russian audience ("How many women, in an enlightened community devoid of superstitious taboos and of sex slavery," he asked rhetorically, "would be eager and proud to bear and raise a child of Lenin or of Darwin!"),[37] and the copy that he presented to Stalin was accompanied by a lengthy personal appeal effusive in its praise of Bolshevism and excoriating the racist and class uses of eugenics in capitalist societies. The strategy failed and Muller shortly thereafter found it advisable to leave the Soviet Union by joining a medical unit in Spain. But while his book did not please Stalin, its reception outside of the Soviet Union was enthusiastic. The *Daily Worker* hailed it as a model for Marxist scientists while *Science and Society* requested that Muller become a foreign editor (he accepted).[38] The book was also well thought of outside of leftist circles. I have been able to locate eighteen contemporary reviews from a wide range of sources (representing professional, general interest journals and newspapers, and both Left and establishment perspectives). Of the dozen that expressed an opinion, only one could be characterized as hostile, and most were decidedly positive.[39]

The most striking feature of the reviews is the widespread assumption (on the part of those who had reservations as well as those who were laudatory) that Muller's claims were scientifically unproblematic. The reservations were social, moral, religious, and political; no one doubted that the scheme was realizable. That traits of character and personality were substantially heritable and that the mechanisms involved were well enough understood that these traits could be consciously manipulated, was taken for granted. Some examples:

> He has produced a scheme by which the human race could be radically changed in two or three generations; scientifically, there is scarcely any doubt that it could be done.[40]

> The important and interesting—though to some of us disturbing—reflection is, as Professor Haldane reminds us, that there is nothing in our established biologic or sociologic knowledge to preclude the material realization of most of Professor Muller's speculations before we are many generations older.[41]

> The author of this book is an American geneticist, at present occupying an important position in a research institute in the Soviet Union. Essentially his thesis is that genetics can and should contribute to a vigorous, practical eugenics of the positive sort. . . . There is in this suggestion nothing unreasonable or impractical either technically or socially.[42]

> [A] torrential procession of ideas and suggestions, often Wellsian in daring but compelling by their reasonableness and convincingly practicable nature.[43]

> It is important to note that technically many of the proposals are either possible at present or likely to be rendered possible by a relatively small amount of further research.

> It is certain that Professor Muller's views on the technique of human improvement will not be universally accepted. Some geneticists do not agree that the use of a few chosen sires is the best way of achieving rapid selection. It can be stated, however, that in spite of possible disagreement on some points, the book will be of great interest to anyone concerned with the problems of eugenics, as it is both genetically and technically accurate.[44]

Evidence from numerous other sources confirms what is suggested by the reviews of Muller's book: the genetic determination of mentality and temperament was taken for granted by geneticists of every political stripe. H. S. Jennings thought it beyond dispute "that such matters as dullness, stupidity, and their opposites, various diversities of temperament, and the like, depend on the genes."[45] Few would have dissented from the claim, expressed in an enthusiastic review of Muller's book in the *Journal of the American Medical Association*, "that what has been found true for the fruit fly is surely applicable to man." The reviewer notes further that "the book is an excellent exposition of the extreme hereditarian doctrine as held by most modern geneticists."[46]

Given that doctrine, it is but a short step to the advocacy of eugenics, although it is one that might not be taken as a result of other moral, religious, or political considerations. These considerations did not figure prominently (when they figured at all) in the outlook of geneticists in the 1920s and 1930s. Hence, they came to see themselves as confronted by enemies on all sides: on the one, "extreme environmentalists" such as the Watsonian behavioralists and many Lamarckians; on the other, proponents of conventional, that is, race- and class-biased eugenics. They perceived themselves as defenders of a reasonable but embattled middle ground, upholding the claims of genetics and the potential social uses of genetics against both those who refused to face scientific facts and those who distorted the facts in the service of racism, nationalism, and class prejudice.

Perhaps the best statement of their position is the "Geneticists' Manifesto" of 1939. Written primarily by Muller, and signed by twenty-two other distinguished geneticists, it was issued at the Seventh International Congress of Genetics at Edinburgh in response to a request from Science Service for a reply to the question: "How could the world's population be improved most effectively genetically?" The central point of the statement, whose signatories included Haldane, Hogben, Huxley, Needham, Dahlberg, Dobzhansky, and Waddington, was that the genetic improvement of mankind depends upon a radical change in social conditions. It was, essentially, a summary of *Out of the Night*. According to the signatories:

> The most important genetic objectives, from a social point of view, are the improvement of those genetic characteristics which make (a) for health, (b) for the complex called intelligence, and (c) for those temperamental qualities which favour fellow-feeling and social behavior rather than those (today most esteemed by many) which make for personal "success,"

as success is usually understood at present. A more wide-spread understanding of biological principles will bring with it the realization that much more than the prevention of genetic deterioration is to be sought for, and that the raising of the level of the average of the population nearly to that of the highest now existing in isolated individuals, in regard to physical well being, intelligence and temperamental qualities, is an achievement that would—so far as purely genetic considerations are concerned—be physically possible within a comparatively small number of generations. Thus everyone might look upon "genius," combined of course with stability, as his birthright.[47]

A Note on Race and Class

The "Manifesto" brands as unscientific the "doctrine that good or bad genes are the monopoly of particular peoples or of persons with features of a given kind," and in *Out of the Night,* Muller concludes that selection could not be responsible for any but insignificant differences among classes or races or other groups. Haldane, on the other hand, believed both that races differed in their "proportions of highly-gifted people" and that the lower classes were less well genetically endowed than the upper. Commenting in a letter on Muller's controversial 1932 speech "The Dominance of Economics over Eugenics," Haldane took issue with his contention that if social classes differed genetically, there was as good a case to be made for the genetic superiority of the masses as of the elite. Those most likely to succeed in the present social order, Muller had argued, were those with predatory characters; the honest, the selfless, the social, and those "too intelligent to confine their interests to their personal success were likely to be left behind."[48] Haldane objected that capitalism was dysgenic precisely because the conventional view was true. A capitalist system ensures that the rich, who are innately superior (since the more able and intelligent are likely to succeed), will be outbred by the poor. Only when the economic position of the latter improves will they choose to have fewer children. (Muller remarked, on a copy of the letter he sent to Solomon Levit in Moscow: "Remember that Haldane is supposed to represent the left most wing in English scientific thought.")[49]

Haldane was at that point a socialist, although not yet a Marxist. However, his standpoint on this matter did not change with his political commitments. Even through the Lysenko period, when he was under severe pressure to abandon, or at least moderate his position, he refused to do so. In public as well as in private, he continued to assert that the upper classes were innately more able and intelligent. Whatever one may think of the content of Haldane's views, it is hard not to admire the independence of spirit that led him to argue in *The Modern Quarterly* (the leading journal of orthodox Marxism in Britain) at the height of the Lysenko controversy: "In many countries the poor breed much quicker than the rich, even when allowance is made for their higher death-rate. Thus the valuable genes making for ability, which bring economic success to their possessors, are getting rarer, and the average intelligence of the nation is declining."[50] In 1957, on the occasion of the Karl Pearson centenary celebration, he asserted that: "Pearson and his colleagues were completely right in one respect. Even if, in spite of his predictions, the nation has improved in some measurable directions, it would have improved more if, say, a million children who were born to unskilled labourers had been born to skilled workers, teachers, and the like."[51]

The Collapse of Consensus

Through at least the early 1940s, there existed a near consensus among geneticists on the substantial genetic determination of intellectual, psychological, and moral traits. It seemed to follow naturally that, at a minimum, the less fit elements of the population should be discouraged from breeding and the more fit encouraged. Even H. S. Jennings, although less optimistic than Muller about what could be accomplished in a relatively short span of time by "positive" eugenics and convinced that environmental measures would have a greater short-term effect, nevertheless argued that positive eugenic measures also have their uses, for if they operate slowly, they do work in the same direction as environmental improvements. And like most of those who were dubious about the short-run utility of positive measures, he emphasized the value of negative ones.[52] As the earlier quote from Lancelot Hogben indicates, skepticism about the possibility of rapid selection does not necessarily imply opposition to eugenics. H. G. Wells considered positive measures futile, but he therefore concluded that it was all the more necessary to sterilize "failures."[53] Eden Paul was equally skeptical and more extreme in his proposed solution. Thus it is not surprising to find Jennings asserting that "habitual criminals not be allowed to propagate" and expressing incredulity "that anyone should knowingly advocate continuing the operations of defective genes that produce such frightful results as idiocy or insanity."[54] In Jennings's view, we may not be able to agree on what constitutes fitness or be able to produce more of it (at least in the short run) if we could, but we do know what constitutes unfitness and we can and ought to act so as to substantially eliminate it. In short, geneticists of diverse political perspectives, including some who were considered "environmentalists," agreed that genes were powerful determinants of mental, temperamental, and moral traits and endorsed some kind of eugenics.

This consensus collapsed with amazing rapidity. Two decades after publication of the "Geneticists' Manifesto," only a relative handful of geneticists remained active defenders of the position that it articulated. How can such a rapid and near-total collapse, or at least apparent collapse, of consensus be explained?

It would certainly not be by developments internal to the science during this period. It is sometimes asserted that the discovery of polygenic inheritance, gene-gene interaction, and gene-environment interaction undermined the assumptions on which support for eugenics was based. However, the existence of these processes had been established much too early to serve as explanations of changes in attitudes occurring in the 1950s and 1960s. Indeed, Muller himself had insisted as early as 1911 that a character is usually the product of several or many genes and always emphasized the complexity of the relationship of genes to traits. A more plausible factor was the supposition, during the 1950s, of a much larger proportion of genetic variability in natural populations than had hitherto been suspected. If populations were so rich in genetic diversity, it seemed reasonable to assume that selection was not acting to "purify" them and that diversity must therefore be advantageous. This view (the "balance" school of population genetics) is associated most prominently with Theodosius Dobzhansky, and it was, at least in part, Dobzhansky's favorable assessment of genetic diversity that led to his (and his colleague L. C. Dunn's) polemics with Muller and the "classical" school during the 1950s and 1960s. But their argument, however significant (and deeply felt to the participants), did not concern the genetic determination of intelligence, character, and personality or the desirability of eugenics per se. It was rather a dispute over the goals and methods of a eugenics program, given widely differing assessments of the value of genetic diversity. Dobzhansky and Dunn always assumed a substantial genetic basis for non-physical human differences and both insisted that they were not hostile to a properly conceived eugenics. That Dobzhansky signed the "Geneticists' Manifesto" of 1939 was consistent with his life-long beliefs. As Richard Lewontin has noted:

Both [the balance and the classical] schools are equally "biologist" in that they believed the nature of human society to be strongly influenced by the distribution of genotypes in the species. For Muller, human progress meant enriching the species for a few superior genotypes while for Dobzhansky it means increasing, or at least maintaining, genetic diversity. Neither view admits the possibility that genetic variation is irrelevant to the present and future structure of human institutions, that the unique feature of man's biological nature is that he is not constrained by it.[55]

Hence theirs was an argument within the consensus that we have been exploring. It is not necessary to minimize the distance separating Dobzhansky and Dunn from Muller to insist that for all that they thought of themselves—and are generally perceived by others as being—antagonists in a deep and sometimes bitter scientific-cum-political dispute, in another perspective they stand together outside of the consensus developing among their contemporaries.

The breakdown of the old consensus is rooted in political, and not scientific, events. Or perhaps it would be more accurate to say that the role played by events internal to science was indirect and largely unrelated to the discovery of new facts or the development of new theories. If one asks what accounts for the development of the original "determinist" consensus, the answer seems obvious. The tendency of scientists to push a new theory to the farthest reaches of its domain—and then beyond—is well known. The history of science includes many examples (none more striking than Darwinism) of attempts to generalize theories and extend the range

of their application far beyond the narrow problems that constituted their original domains. This is presumably true also of the
early decades of genetics. How else explain why scientists of
every possible political persuasion—conservatives, liberals, Marx-
ian and non-Marxian socialists—share a common commitment
to what we would today call biological determinism and sympathy for some kind of eugenics? Geneticists of the early decades
of this century agreed on nothing except the proposition that the
salvation of mankind was to some extent bound up with hereditary improvement. Whatever their broader politics, they were
all genetic imperialists.

But it is also typical that after the first extreme phase—the
period in which attempts are made to generalize the theory, to
make of it a new world view, to extend the range of its scientific
and social applications—that a reaction occurs. Those routed in
other fields begin to regroup and to defend themselves; even the
imperialists begin to have doubts. This natural process of retreat
following (over) expansion would almost certainly have undermined consensus among geneticists even in the absence of the
momentous social forces that ultimately swamped it. As it was,
the consequences of Hitler's rise to power in Germany and Stalin's
in the Soviet Union were enough to throw into disrepute, at least
temporarily, the assumptions shared by nearly all geneticists until
the mid-1940s.

Biological explanations of non-physical human differences rapidly lost favor in the general revulsion toward the uses for which
they had been employed by the Nazis. Left geneticists were of
course affected by that development, but also by events specific to
socialists. The 1940s witnessed the rapid intensification of pressures both from without and from within their own camp, pressures that had threatened the existence of a socialist eugenics from
its inception. In the 1920s and 1930s, scientists such as Muller,
Haldane, Huxley, and Needham had struggled to disassociate their
program from the racially and class distorted eugenics of the Right
while at the same time battling the environmentalism of the Left.
During the 1940s, assaults from the Right and Left escalated in
intensity; racial and class prejudice gave way to Fascism and environmentalism to the reign of T. D. Lysenko. What had always
been a precarious middle ground, defended against the environ-

mentalism of their political allies and the racial and class prejudices of other eugenicists, simply collapsed.

Moreover, even those Left scientists who broke with the Soviet Union and who therefore remained free to continue asserting the relevance of genetics to society, escaped one horn of a dilemma only to impale themselves on another. For the claim of the Left geneticists had always been that the biological improvement of mankind presupposed the transformation of social relationships. It was only in a society providing equal opportunities to all its members that a eugenics program would be defensible. In the early 1930s, most believed both that the Soviet Union was or was becoming such a society and also that it would provide a model for Western industrial states, then caught in the grip of the Great Depression. Faced with the persistence of capitalism in the West, those who concluded that the Soviet experiment had failed were forced by the logic of their position to abandon their eugenical hopes. But neither Muller nor Haldane nor Huxley, all of whom continued to argue the case for eugenics into the 1960s, ever confronted the apparent inconsistency of their position. Indeed, what is most striking about the Left geneticists in general—including those such as Dobzhansky and Dunn, whose views were considerably more moderate—is how little they changed over the years. Circumstances changed, shifting the ground from under their position, but their own views were little affected. Their cause failed, not as the result of desertions from the ranks, but from the inability to win new recruits.

Virtually all of the Left geneticists whose views were formed in the first three decades of the century died believing in a link between biological and social progress. Their students, coming to intellectual maturity in a radically different social climate, either did not agree or, in a social climate inhospitable to determinism, were unwilling to defend that position. The appearance of sociobiology probably signifies a fading of the bitter memories surrounding the events of the 1940s. As those memories recede, it would not be surprising to witness the re-emergence of a doctrine that was never defeated in the scientific arena but rather submerged by political and social events. From the late 1940s to the early 1970s, it has been, perhaps, a viewpoint latent among scientists only requiring another change in the social climate to prompt its expression.

—————————————————— Notes ——————————————————

1. Diane B. Paul, "Marxism, Darwinism, and the Theory of Two Sciences," *Marxist Perspectives* 2 (Spring 1979): 116–43; *idem*, "'In the Interests of Civilization': Marxist Views of Race and Culture in the Nineteenth Century," *Journal of the History of Ideas* 32 (January 1981): 115–38; *idem*, "A War on Two Fronts: J. B. S. Haldane and the Response to Lysenkoism in Britain," *Journal of the History of Biology* 16 (Spring 1983): 1–37.

2. Daniel J. Kevles, *In the Name of Eugenics* (New York: Alfred A. Knopf, 1985).

3. Mark Haller, *Eugenics: Hereditarian Attitudes in American Thought* (1963); Kenneth Ludmerer, *Genetics and American Society* (1973).

4. J. B. S. Haldane, "Darwin on Slavery," *Daily Worker* (London), November 14, 1949. In *The Inequality of Man and Other Essays* (London: Chatto and Windus, 1932), Haldane asserted that "The test of the devotion of the Union of Soviet Socialist Republics to science will, I think, come when the accumulation of the results of human genetics, demonstrating what I believe to be the fact of innate human inequality, becomes important" (p. 137). After he joined the British Communist Party, this claim would return to haunt him. The tension between Haldane's political and scientific views is explored in Diane B. Paul, "A War on Two Fronts."

5. For expressions of Haldane's late views, see his speech published in *Karl Pearson: Centenary Celebration*, May 13, 1957 (London: privately issued by the Biometrika Trustees, 1958); "The Proper Application of the Knowledge of Human Genetics," in M. Goldsmith and A. Mackay, eds., *The Science of Science* (London: Souvenir Press, 1964), pp. 150–56; and "The Implications of Genetics for Human Society," in *Genetics Today: Proceedings of the XI International Congress of Genetics, The Hague, September, 1963* (New York: Macmillan, 1965), vol. 2, pp. xci–xcii.

6. "Negative" eugenics is oriented toward the reduction or elimination of unfavorable characteristics in a population; "positive" eugenics, toward the increase of favorable characteristics. An example of the former would be sterilization of the "feebleminded"; of the latter, artificial insemination of women with the sperm of Nobel Prize winners.

7. George Bernard Shaw, *Sociological Papers* (London: Macmillan, 1905), pp. 74–75. Shaw was a lecturer of the Eugenics Education Society; Eden Paul, "Eugenics, Birth Control, and Socialism," in Eden Paul and Cedar Paul, eds., *Population and Birth-Control: A Symposium* (New York, 1917), p. 139; H. J. Laski, "The Scope of Eugenics," *Westminster Review* 174 (1910): 34. See also Havelock Ellis, "The Sterilization of the Unfit," *Eugenics Review*, October 1909, pp. 203–206; *idem, The Problem of Race Regeneration* (London: Cassell, 1911); *idem, The Task of Social Hygiene*

(London: Constable, 1912); H. G. Wells, *Sociological Papers* (London: Macmillan, 1905), pp. 58–60; *idem, A Modern Utopia* (New York: Charles Scribner's Sons, 1905), esp. pp. 175–86; Sidney Webb, *The Decline in the Birth-Rate* (London: Fabian Society, 1907); *idem,* "Eugenics and the Poor Law: The Minority Report," *Eugenics Review* 2 (1910–1911): 233–41; Sidney Webb and Beatrice Webb, *The Prevention of Destitution* (London: Longmans, 1911); Graham Wallas, *The Great Society: A Psychological Analysis* (New York: Macmillan, 1914), esp. pp. 55–56; Paul Kammerer, *The Inheritance of Acquired Characteristics,* trans. A. Paul Maerker-Branden (New York: Boni and Liveright, 1924), Part B: *Eugenical Part,* esp. Chap. 53 "Productive Eugenics"; Edward Aveling, *Progress* 2 (1883): 210–17; *idem, Darwinism and Small Families* (London: printed by Annie Besant and Charles Bradlaugh, 1882); Karl Pearson, *The Ethic of Free Thought* (London: Adam and Charles Black, 1901); *idem, The Problems of Practical Eugenics* (London: Dulau, 1912); Edward Bellamy, *Looking Backward* (New York: Signet, 1888; rpt. 1960), esp. pp. 179–81; Herbert Brewer, "Eutelegenesis," *Eugenics Review* 27 (1935): 121–26; Julian S. Huxley, "Eugenics and Society," *Eugenics Review* 28 (1936): 11–31; *idem, Memories I* (London: Allen and Unwin, 1970); H. J. Muller, "The Dominance of Economics over Eugenics," *Birth Control Review* 16 (1932): 236–38; *idem, Out of the Night* (New York: Vanguard Press, 1935; English edition, 1936).

8. Some historians have questioned the value of the term "social Darwinism." Thus, Robert Bannister has argued that the label was invented essentially to slander the people it was applied to and in fact accurately describes the views of very few thinkers. See his *Social Darwinism: Science and Myth in Anglo-American Social Thought* (Philadelphia, 1979).

9. In his review of the literature on eugenics, Lindsay Farrall writes: "Today, eugenics tends to be dismissed as a pseudo-science or as a species of Social Darwinism which received support from political reactionaries or racial bigots. It is true that eugenics was used to support reactionary and racist views; but eugenic ideas and the eugenics movement were much too complex and significant to allow simplistic historical judgements to go unchallenged." See "The History of Eugenics: A Bibliographical Review," *Annals of Science* 36 (1979): 111. This essay represents one attempt to call into question those "simplistic historical judgments." Others who in one or another way have already done so are: Jonathan Harwood, "Nature, Nurture and Politics: A Critique of the Conventional Wisdom," in J. V. Smith and D. Hamilton, eds., *The Meritocratic Intellect* (Aberdeen, 1980), pp. 115–28; Loren Graham, "Science and Values: The Eugenics Movement in Germany and Russia in the 1920s," *American Historical Review* 82 (1977): 1133–64; Michael Freeden, "Eugenics and Progressive Thought: A Study in Ideological Affinity," *The Historical Journal* 22 (1979): 645–71; and (in

respect to Fabianism) Donald MacKenzie, "Eugenics in Britain," *Social Studies of Science* 6 (1976): 449–532; *idem*, "Karl Pearson and the Professional Middle Classes," *Annals of Science* 36 (1979): 125–36. G. R. Searle has taken issue with MacKenzie's thesis on the affinity of Fabianism with eugenics in "Eugenics and Class," in Charles Webster, ed., *Biology, Medicine and Society 1840–1940* (Cambridge, 1981), pp. 217–42, arguing that we must avoid the "absurd situation" in which "the 'eugenist' label is going to be placed around the neck of nearly every major political and social thinker" (p. 239). On the relationship of eugenics to the women's movement, see Linda Gordon, *Woman's Body, Woman's Right: A Social History of Birth Control in America* (New York, 1976), esp. pp. 116–35; David M. Kennedy, *Birth Control in America: The Career of Margaret Sanger* (New Haven, 1970), esp. pp. 114–22. (The slogan of Sanger's American Birth Control League was "To Breed a Race of Thoroughbreds.") R. A. Soloway, *Birth Control and the Population Question in England, 1877–1930* (Chapel Hill: University of North Carolina Press, 1982). This relationship is a focus of Daniel J. Kevles, *In the Name of Eugenics*.

10. In 1935, the editors of the *Eugenics Review* noted: "It almost seems as if geneticists in this country will have to add Russian to their already formidable linguistic equipment." (See "Notes of the Quarter," *Eugenics Review* 27 [1935]: 188.) A year later they also remarked: "In recent issues of the *Review* we have drawn attention to signs of increasing interest and sympathy with eugenics" on the part of persons and parties belonging to the political Left. See "Progressive Parties and Eugenics," *Eugenics Review* 28 (1936): 296.

11. "Eugenics and the Poor Law," p. 237.

12. *Genetic Principles in Medicine and Social Science* (London: Williams and Norgate, 1931), p. 207.

13. The phrase is taken from "Social Biology and Population Improvement," *Nature* 144 (1939): 521. This article was popularly known as "the Geneticists' Manifesto."

14. The liberal critique is exemplified by J. A. Hobson, "Race Eugenics as a Policy," in *Free Thought in the Social Sciences* (New York: Macmillan, 1926), pp. 200–21. While Leonard Hobhouse rejected "positive" eugenics, he decried reproduction by "men and women who are not capable of independent existence but who continually drift to the gaol or the workhouse, who are fertile, and whose condition is asserted to be hereditary in a marked degree" (*Social Evolution and Political Theory* [New York: Columbia University Press, 1911], pp. 45–46). Also: "We are dealing with people who are not capable of guiding their own lives and who should for their own sake be under tutelage and we are entitled to impose our own conditions of this tutelage having the general welfare of society in view" (*ibid.*, p. 46). The Pope indirectly repudiated eugenics in the 1930 Encyclical, "On

Christian Marriage," and later statements. See Pius XI, "On Christian Marriage." The English Translation (New York: Barry Vail, 1931).

15. G. K. Chesterton, "To the Reader," in *Eugenics and Other Evils* (London: Cassell, 1922), p. vi.

16. J. B. S. Haldane, *Heredity and Politics* (New York: W. W. Norton, 1938), p. 8.

17. Donald MacKenzie, "Eugenics in Britain," p. 520. MacKenzie draws a sharp distinction between Marxism and Fabianism.

18. Quoted in J. B. S. Haldane, *Heredity and Politics,* p. 14.

19. As Hogben wrote: "Negative eugenics is simply the adoption of a national minimum of parenthood, an extension of the principle of national minima familiarized in the writings of Sidney and Beatrice Webb. It is thus essentially *en rapport* with the social theory of the collectivist movements." *Genetic Principles,* p. 210.

20. See Gary Werskey, *The Visible College: A Collective Biography of British Scientific Socialists of the 1930s* (New York: Holt, Rinehart, and Winston, 1978). Werskey's findings are elegantly summarized by Martin Green in "The Visible College in British Science," *The American Scholar* 47 (1977/78): 105–17.

21. Karl Pearson, "The Moral Basis of Socialism," in *The Ethic of Free Thought* (London, 1901; first published in 1887), pp. 307–308.

22. Garland Allen, *Thomas Hunt Morgan: The Man and His Science* (Princeton: Princeton University Press, 1978), p. 228.

23. Brewer, "Eutelegenesis," pp. 121–26.

24. Letter of Herbert Brewer to Joseph Needham (1936; otherwise undated), Joseph Needham Papers, Section 5 (Part II) *Social Biology,* 1936–1946, Cambridge University Library. Haldane was later to become skeptical of such proposals. Shortly before his death he wrote: "I fully agree with Muller that in so far as artificial insemination is practiced, the donors should be chosen to be as desirable as possible genetically. I am more skeptical that this or any other scheme which we can devise at present would greatly improve the genetical make-up of our species. . . . This is not to say that very great improvement is not possible. However, I do not think we know much more about how to bring it about than Galileo or Newton knew about how to fly." Haldane, *Genetics Today,* p. xcvi.

25. Muller, *Out of the Night,* p. 90.

26. *Ibid.,* chapter 6, *passim.*

27. *Ibid.,* p. 105. Also: "On the part of a host of intelligent women, therefore, there is a growing mass strike against child-bearing. May the strike prosper until the dire, age-old grievances have been removed" (*ibid.,* p. 104). At the birth of their son in 1924 Muller's first wife was fired from her teaching position in the mathematics department at the University of Texas because the department "felt that a mother could not give full attention to classroom

duties and remain a good mother." See Elof Axel Carlson, *Genes, Radiation, and Society: The Life and Work of H. J. Muller* (Ithaca: Cornell University Press, 1981), p. 133.

28. *Ibid.*, pp. 103–108.

29. *Ibid.*, p. 108.

30. References to *Daedalus* appear on pp. 74 and 110. J. B. S. Haldane, *Daedalus: On Science and the Future* (London: K. Paul, Trench, Trubner, 1924), pp. 63–68; J. D. Bernal, *The World, The Flesh, and the Devil* (Bloomington: Indiana University Press, 1969; orig. ed. 1928), pp. 30, 36.

31. Muller, *Out of the Night*, p. 118.

32. *Ibid.*, p. 113.

33. *Ibid.*, p. 114.

34. *Ibid.*, pp. 114–15.

35. Graham, "Science and Values"; The Reminiscences of Theodosius Dobzhansky, Columbia Oral History Collection.

36. He was apparently advised in this by Solomon Levit. See Carlson, *Genes, Radiation, and Society*, p. 228. For Stalin's reaction, see p. 233.

37. Muller, *Out of the Night*, p. 122.

38. Letter of David Ramsey to Muller, August 28, 1936, asking him to contribute to the first issue and to serve as a foreign editor. Muller Collection, Lilly Library, Indiana University.

39. The hostile review was by A. J. Carlson, *American Journal of Sociology* 42 (1936): 134. Reviews other than those cited in notes 40–44 are: Waldemar Kaempffert, *New York Times*, March 15, 1936, p. 4; Ray Erwin Baber, *American Sociological Review* 1 (1936): 533; P. W. Whitney, "Communist Eugenics," *Journal of Heredity* 27 (1936): 132–35; J. L. Stocks, *Manchester Guardian*, June 9, 1936, p. 7; Harold Ward, *New Republic* 86 (1936): 284; *Survey: Journal of Social Work* 72 (1936): 159; *Adelphi* 13 (1936): 192; *Booklist* 32 (1936): 383; Julian Huxley, "Marxist Eugenics," *Eugenics Review* 1 (1936): 66–68; *Journal of the American Medical Association* 107 (1936): 68; *Scientific American* 154 (1936): 230; *Quarterly Review of Biology* 11 (1936): 348.

40. C. P. Snow, "Revolution in Ourselves," *Spectator* 157 (1936): 64. Also: "there is no doubt that the supreme abilities, which make a man useful in the world, are a property of the genes. When genetics is more universally understood it will not be easy for us all to escape the consequences which that truth brings."

41. *Times Literary Supplement*, June 20, 1936, p. 526.

42. Percy Dawson, *Unity*, May 18, 1936, p. 115.

43. "P. J." (unidentified author), *Science Progress* 31 (1937): 790.

44. K. Mather, *Nature* 138 (1936): 228.

45. H. S. Jennings, *The Biological Basis of Human Nature* (New York: W. W. Norton, 1930), p. 157.

46. [Review of H. J. Muller, *Out of the Night*], *Journal of the American Medical Association* 107 (1936): 68.

47. "Social Biology and Population Improvement," *Nature* 144 (1939): 521.

48. Muller, *Out of the Night,* pp. 89–90.

49. Letter of Haldane to Muller, July 29, 1932. Muller Collection.

50. J. B. S. Haldane, "Biology and Marxism," *Modern Quarterly* 3 (1948): 9.

51. Haldane, speech in *Karl Pearson: Centenary Celebration.*

52. Jennings, *Biological Basis,* esp. chapters 4 and 6.

53. Wells, *Sociological Papers,* p. 60. "It is in the sterilization of failures, and not in the selection of successes for breeding, that the possibility of an improvement of the human stock lies."

54. Jennings, *Biological Basis,* p. 238; *The Reminiscences of Theodosius Dobzhansky; The Reminiscences of L. C. Dunn,* Columbia Oral History Collection, *passim.*

55. R. C. Lewontin, *The Genetic Basis of Evolutionary Change* (New York: Columbia University Press, 1974), p. 31.

3

THE NINE LIVES OF DISCREDITED DATA

New Preface

In 1985, I published an essay on discussions of the genetics of intelligence in genetics textbooks, describing the situation as grim.[1] Discussions of the concept of heritability were confused in the extreme. Even more surprising, the Cyril Burt scandal of the mid-1970s had only trivially affected the content of texts. While Burt's name had disappeared—except as an example of fraud in science—his data were still reported. But the most astonishing feature of the textbooks was their similarity. The same data were cited in support of the same conclusions, often in practically the same words, in text after text.

In an effort to understand this similarity, I interviewed many textbook editors and publishers. The article that resulted from this research generated a raft of mail from other editors, textbook authors, industry analysts, and scholars, some with stories that were even more startling.[2] Of those brought to my attention, that of John B. Watson and Rosalie Rayner's "Little Albert" experiment seemed particularly instructive. I teamed with Arthur Blumenthal, a historian of psychology, to explore the fate of this classic "environmentalist" fable in psychology textbooks. The results, described in the epilogue to this chapter, were strikingly similar.[3]

I have also incorporated some material from a short related piece, "The Market as Censor," and added notes where I wished to acknowledge recent developments or points brought to my attention since the article originally appeared in *The Sciences*.[4] Since data on college textbook mergers, production costs, sales, and the like constantly change (and since the specifics do not matter for the argument), I have not tried to update these details. The forces and trends identified in the essay have, if anything, intensified since the original article was published.[5]

One of the lessons today's undergraduate science majors may glean from their genetics textbooks is that differences in intelligence, as measured by I.Q., are due primarily to differences in genes. Students consulting John B. Jenkins's *Human Genetics,* a current best-seller, will learn that "the genotype has a greater influence on I.Q. than do environmental factors."[6] And those studying H. Eldon Sutton's *Introduction to Human Genetics* will be given to understand that I.Q. variations are "largely genetic in whites."[7]

Fifteen years ago, few geneticists would have argued with such assertions. That I.Q. is seventy to eighty percent heritable seemed indisputable in the light of experiments performed by the English psychometrician Cyril Burt on groups of identical twins. In five studies published between 1955 and 1966, Burt and his collaborators reported that the I.Q. scores of identical twins were always closely matched, whether the twins had grown up together or apart. His results were authoritative—no other investigator had claimed such success at tracking down twins who had been separated at birth and reared in different environments—and they seemed conclusive. There was just one problem: Burt's impressive findings were fraudulent.

Suspicions were aroused in 1972 (a year after Burt's death), when the psychologist Leon Kamin noted that Burt's I.Q. correlations (0.771 for twins raised separately, 0.944 for those raised together) had remained constant throughout various studies involving different numbers of subjects—nothing short of a statistical miracle.[8] Then, in 1976, Oliver Gillie, a medical correspondent for the *Sunday Times* (London), reported evidence that Burt had invented his ostensible research collaborators, "J. Conway and M. Howard," and had fabricated some of his data.[9] These charges were not proved until 1979, when Burt's biographer, Leslie Hearnshaw, confirmed by examining personal diaries that Burt had never conducted many of the studies he reported. But no one had doubted since the mid-1970s that, whether through incompetence or chicanery, Burt's work was tainted.[10]

Authors of genetics textbooks responded to these events in a curious way: they stopped citing Burt as an authority, but many continued to cite his results. In a study of twenty-eight texts published between 1978 and 1984, I found that most of the nineteen discussing the heritability of I.Q. assert that it is high. As evidence, eleven of these texts cite a review article, published in the journal *Science* in 1963, in which L. Erlenmyer-Kimling and Lissy F. Jarvik incorporated results from fifty-two early studies into a figure indicating a strong inverse correlation between I.Q. variations and degrees of genetic relatedness.[11] Had the authors of these textbooks read the review article closely, they would have noticed that it included Cyril Burt's results. Yet most (eight out of the eleven) went so far as to reproduce the figure that accompanied it.

How could so many authors be so thoroughly out of touch? The answer lies in the dramatic changes that have taken place over the past quarter-century in the way textbooks are published. The repetition in text after text of discredited data is part of a larger trend toward greater emphasis on packaging and less concern with content. Today's textbooks are thicker, slicker, more elaborate, and more expensive than they used to be. They are also more alike. Indeed, many are virtual clones, both stylistic and substantive, of a market leader. These trends are not unique to genetics texts: in fact, cribbing—authors' borrowing liberally from other textbooks—is widespread. And as the bizarre durability of Cyril Burt's data makes clear, the practice can have pernicious, if unintended, consequences.

As recently as the 1960s, textbooks tended to be idiosyncratic, reflecting the author's own approach in both style and substance. Three introductory genetics texts that led the field at the end of the decade—Adrian Srb, Ray Owen, and Robert Edgar's *General Genetics*; Monroe Strickberger's *Genetics*; and Eldon Gardner's *Principles of Genetics*—varied considerably in organization, emphasis, and tone. The same could be said of texts in other fields, including such classics at Robert Winthrop White's *Abnormal Psychology*, P. A. M. Dirac's *Principles of Quantum Mechanics*, Eugene P. Odum's

Fundamentals of Ecology, and Linus Pauling's *Nature of the Chemical Bond*. Their singularity was not surprising, since authors wrote texts mainly to impress their stamp on a field. "When I first came into the [textbook industry]," David P. Amerman, a marketing director at Prentice-Hall, recalls, "the way you published a book was to find an academic with a reputation and hope he could write."[12] If he could not, editors were inclined to preserve the author's voice, even at the expense of readability.

The trend toward homogenization began with the enrollment surge of the 1960s. During that decade, the number of undergraduates in U.S. colleges more than doubled. The most rapid rise occurred in state schools, particularly in two-year community colleges, where nationwide enrollment rose from fewer than half a million in 1960 to more than two million in 1970. The expansion opened a whole new market, which textbook publishers moved aggressively to exploit. Two-year schools became a mainstay of the industry, and remain so, enrolling more than forty percent of all undergraduates.

But community colleges demanded a new sort of textbook. In many ways, these institutions were more like high schools than traditional four-year colleges. Faculty members were not expected to do research and so were given heavy teaching loads: four, five, even six courses a semester, sometimes covering every subfield of a discipline. Since instructors were often not well equipped to handle such a wide range of subjects (few had PhDs and many were part-time), they looked for texts that came with teaching manuals and ready-made tests. Indeed, some community college instructors were former high school teachers who had come to expect such satellite materials.

Instructors also demanded simpler texts, because their students had poorer reading skills, on average, than students at four-year schools. Some community colleges even required that books be written at a tenth grade reading level, as defined by such standard tests as SMOG, Flesch, or the Fry Graph (which measure number of syllables, length of sentences, and familiarity of words). Since most publishers yearned to capture as wide a market as possible, they adjusted the reading levels of their texts—and the nature of supplemental materials—to community college standards.

Meanwhile, the changing demands and increasing volume of the college textbook market attracted a new kind of publisher—one with a heightened concern for the bottom line. A number of conglomerates entered the market, including ITT (which acquired G. K. Hall and Bobbs-Merrill), IBM (which bought Science Research Associates), CBS (which acquired Holt, Rinehart and Winston and others), RCA (one-time owner of Random House), Raytheon Company (which purchased D. C. Heath and Company), Bell and Howell (which bought Charles E. Merrill), and Xerox Corporation (which acquired Ginn and Company, then sold it to Gulf and Western, owner of Prentice-Hall and Allyn and Bacon).

The new players were prepared to invest huge sums in texts, and this had the effect of reducing competition by raising the costs of production and driving smaller presses into specialized niches or out of the market altogether. By 1975, the ten largest college publishers controlled seventy-five percent of total sales, with the top four firms controlling forty percent.[13]

Revision cycles were speeded up, making texts quickly obsolete. Publishers dressed up their books with photographs, art work, and full-color figures; packaged them with such accessories as instructors' manuals, slides (with accompanying lecture notes), and tutorial programs on floppy disks; and even offered subsidies for the purchase of educational films. In time, lecture outlines, alternative course syllabi, slides and transparencies, and computer simulations were added. Large banks of test questions, sold with the texts, were offered in a variety of formats: on floppy disks, formatted for the instructor's personal computer; on magnetic tapes, for use on the campus mainframe; or as separately bound booklets. With these test banks, instructors could generate tests on specific chapters, topics, or course objectives, which some publishers offered to print. The preface to *Psychology: An Introduction* (a current best-seller) assures teachers that "test preparation and typing can be obtained within 24 hours through the Prentice-Hall phone-in testing service."[14]

Such "bells and whistles," as one textbook editor describes them, are expensive. Professionals in test construction, for instance, charge three to five dollars for each of the one thousand to two thousand questions in a typical test bank. Technical illustrators may charge as much as four hundred dollars for a single drawing. And quarter-page photographs, of which there are often hundreds in a basic text, cost as much as two hundred and fifty dollars each, just for permission to reproduce. As a result, publishers came to spend increasing amounts of time and money on packaging. Today, the prevailing belief is that a basic science, social science, or business text that does not include the standard satellite materials will fail—regardless of its other virtues—since many instructors look first at the supplements and only later at the text itself.

Some publishers also offered material inducements in return for textbook adoptions. These include subsidies toward the purchase of educational materials, such as film rentals, payments to teachers of large sections or members of adoption committees for "reviewing" the text, where it is understood that the decision to adopt, not the comments, is important, provision of extra complimentary copies for resale on the used book market, offers to both departments and individuals of equipment or direct cash payments thinly disguised as "grants" for educational purposes in return for textbook adoptions.[15] These developments are essentially invisible to the students, who pay the costs in inflated text prices. But just as the customers for prescription drugs are physicians, not patients, the real customers for texts are instructors, not students.[16] And instructors have little awareness of, or sensitivity to, price differentials.[17]

As the market grew and textbooks changed, some publishers also began to look for a different kind of author. They became less interested in a writer's scientific expertise and more concerned with the ability to reach a mass audience. Hence, many publishers stopped recruiting authors from prestigious universities—where professors may not have taught introductory courses in years and were more prone to write for their peers than for students—and began to look for successful teachers of large classes at state schools. But, in the end, this development was probably less significant than changes in the authors' own motivations.

Some textbooks are, of course, still written out of authors' beliefs that they have something important to offer, and these authors have no incentive to copy other books; indeed, it would defeat their purpose. But some editors say that a new kind of writer has emerged: one motivated more by potential profit than by the desire to leave an intellectual legacy. Writing texts had never conferred great prestige and rarely earned their authors much money. But with the enrollment surge of the 1960s, it became a plausible route to wealth. As a result, the field began to attract authors who have little emotional involvement with the text and few ideas of their own—authors who draw inspiration from editors and, especially, from other textbooks.

The incentive to borrow from other texts is heightened by the need to cover an expanding number of topics. Since the mid-1960s, textbooks have increased in both page surface area and number. Thus, biology texts have increased in length by about two-thirds (most are now between eight hundred and twelve hundred pages long), and the average length of psychology textbooks has grown from fewer than five hundred pages to more than seven hundred. This is partly due to the expansion of knowledge—many of the topics in current texts, such as genetic engineering or sociobiology, scarcely existed more than twenty years ago—and the increased use of graphics, but it is also partly the result of marketing considerations. Publishers trying to capture the largest possible market are loathe to omit anyone's pet topic. Professors asked to review manuscripts often agree that the text is too long but may not agree on what should be cut. Hence, the safest policy is to leave everything in, and textbooks grow without evidence that students are actually reading more pages.

Of the multitude of topics covered in contemporary textbooks, the author is likely to have experience in only a few. One way to master the unfamiliar topics, of course, is to read the professional literature—to comb through specialized monographs and journals. Some authors do this. But it is much easier to borrow predigested material from other textbooks. And with so many textbooks currently in print—more than a hundred in introductory psychology alone—authors who crib can feel secure that their sources will not be easily identified.

Even when authors want to be original, publishers may pressure them to conform. At times this pressure is quite overt: the publisher explicitly sets out to mimic the style and content of the most successful text in the field. In a 1974 lawsuit, Harper and Row charged the Meredith Corporation with plagiarizing its developmental psychology text *Child Development and Personality*, which at the time was enjoying approximately a thirty percent market share. The suit unearthed internal memoranda indicating that Meredith had hired freelance writers, many having no background in psychology, and had provided them with detailed chapter outlines of the Harper and Row text, on which they were to base their drafts. (These draft chapters were to be edited by a well-known psychologist, the official "author" of the text.) One memorandum even warned writers to "resist the temptation to impose your own view of the subject matter; the model [the Harper and Row text] and the marketing report are the arbiters combined with your own common sense."[18]

Such extensive copying of a single text is unusual. What is not unusual, however, is fear of deviating from the mainstream—from textbook formulas that have already proved successful. It is not uncommon for a press to invest as much as half a million dollars in a single text. To protect that investment, the publisher relies heavily on the results of market research and manuscript reviews to ensure that the product is salable. These results, as it turns out, almost inevitably prod the publisher to produce a textbook that resembles all others in the field.

Virtually all publishers use the same forms of market research—principally questionnaires that ask potential adopters of a text how much emphasis various topics should receive, in what sequence they should appear, and how the book in question compares with others. The research typically indicates that most college teachers will resist any change in a textbook that necessitates revising their lecture notes. That research reinforces a lesson that editors learn early in their careers; about seventy-five percent of college editors either began in sales or held sales or marketing positions before becoming editors.[19] Their personal contacts and experience trying to sell innovative texts teach them that college teachers are profoundly conservative. Editors know that "anything that increases work for the professor is not good."[20] In a 1971 lawsuit involving the alleged plagiarism of Campbell R. McConnell's *Economics: Principles, Problems and Policies*, the judge summarized one of the defendant's successful arguments as follows:

> Economics professors, who shape the market, desire texts to which their own class notes can be adapted. Their notes, in turn, are the products of long familiarity with what might be described as "Samuelson methodology." These professors are presumptively unwilling to effect a reorganization of their own notes merely to satisfy the whim of a new textbook writer.[21]

If the guidelines an author receives for writing his book are based largely on market research, the even more crucial editing of the text depends largely on manuscript reviews. Two types of reviewers are used: experts, chosen for their ability to judge the accuracy of the text, and "market knowledgeable" reviewers, selected not for their expertise but for their preferences as consumers.

The larger and more competitive the market for a text, the greater the dependence on market-knowledgeable reviewers; for an introductory text in a field such as psychology, market-knowledgeable reviewers often outnumber experts by two to one. Such reviewers naturally reflect the market's conservatism, and when they dislike what is original in a new book's organization or approach, the editor often responds by encouraging the author to "study" other texts. Thus, the whole process of textbook development conspires to wash out any substantive innovation, even in books that were originally attractive because they appeared to offer something new. But publishers must do *something* to distinguish their texts from the dozens of others on the market, so while meaningful innovations are eliminated, novelty is introduced in the externals—the color illustrations, teaching manuals, lecture slides, and test banks. Originality is thus restricted to areas in which it is trivial, and it becomes little more than a strategy for marketing the same old book under a new author's name.

The pressures that have produced so many meaningless variations on standard textbooks are, if anything, increasing with hard times in the industry. College enrollment stabilized around 1981 and then began to decline while a growing used book industry has added to the strains on publishing houses.[22] As a consequence, the number of hardbound texts sold has been steadily declining. The conglomerates that bought out so many textbook publishers twenty years ago, with visions of virtually risk-free profit, have now begun to sell them. Textbook publishing, in short, has become an intensely competitive business.

The competition might have inspired greater innovation in the writing of texts. Instead, it has created a situation in which textbooks are being produced and sold like toothpaste. In the *Book Publishing Annual of 1984*, industry analyst Thomas W. Gornick summed up the new ethic with his prediction that future textbooks will have "more elaborate designs and greater use of color. . . . The ancillary packages will become more comprehensive, resembling the elementary-high school materials, and more costly. . . . New, more aggressive marketing plans will be needed just to maintain a company's position. The quality of marketing will make the difference."[23]

One could argue that these developments are really no cause for alarm. After all, not every textbook published before 1970 was a model of wit, clarity, and scholarship. Some of the old, idiosyncratic texts were genuinely inspiring to students, but others were simply exercises in self-indulgence: poorly written, lightly edited, and unintelligible to anyone but a specialist. The prose in today's homogenized primers may be bland, but in most cases it is clear. And there is no denying that the lavish use of photographs, figures, and illustrations has made textbooks more engaging. Nor is their substantive similarity a bad thing per se. The purpose of an introductory text is to summarize the central facts and theories of a discipline, not to break new ground or convey novel insights. Books covering the same material are bound to be similar. So what is the problem?

If the leading texts were ideal, there would be no problem. But when the models are flawed, imitating them stifles development of better ones. And to the extent that imitation consists of cribbing information or insights, it guarantees that textbooks will become less reliable as a field advances. Authors working from the professional literature are not likely to fill a text with dated ideas and discredited data. But an author drawing from existing textbooks, even good ones, has no way of knowing whether they are describing the current state of a discipline. Rather than discard damaged remnants from the past, such as Cyril Burt's studies of I.Q. heritability, he gives them a new air of authority.

It is doubtful that the authors still publishing such data are trying to mislead their readers; more likely, they are simply playing by the industry's new rules—modelling their textbooks on others and ignoring the literature they claim to be summarizing. In the genetics texts, this is often obvious from the manner in which the literature is treated. The most frequently cited evidence for the influence of genes on intellectual performance (many of the texts cite no sources at all, even when reproducing charts and graphs from other works) is the figure that accompanied the 1963 review article by Erlenmeyer-Kimling and Jarvik. Anyone who took a close look at that article or read the numerous critiques of it in *Science* and elsewhere would see that (among other shortcomings) it incorporated Burt's data. But authors relying on other textbooks are not privy to such insights. Some, because they are familiar with the Cyril Burt scandal, end up actually denouncing the same data they are reporting. Jenkins's *Human Genetics*, for example—after saying the figure accompanying the 1963 article points out clearly the strength of the genetic component of I.Q.—notes that Burt's data were "manipulated," resulting in their exclusion from current reviews.[24] Similarly, Robert H. Tamarin's *Principles of Genetics* (1982)—after citing the 1963 figure as evidence that "the measured heritability of I.Q. is relatively high"—spends two pages detailing "the case against Sir Cyril Burt."[25] Still another text, Anna Pai and Helen Marcus-Roberts's *Genetics: Its Concepts and Implications* (1981), documents the link between intellectual and genetic variation with a diagram showing test scores of one hundred and twenty-two twin pairs, fifty-three of which are necessarily Burt's. Yet it, too, recounts the Cyril Burt scandal, and even refers readers to studies by Kamin and others who helped discredit Burt's data.[26] Students exploring these suggested readings would be hopelessly baffled by the contradictions between the text and its ostensible sources. But no one expects students to be so enterprising. References are essentially decorative; one editor at a major publishing house calls them "window dressing."[27]

It would be a mistake to presume that genetics texts alone are propagating this sort of nonsense; the practices that generate it are evident throughout the industry. But reliable textbooks are especially important, and shoddy ones particularly invidious, in the sciences. For whereas humanities professors often assemble reading lists from current paperbacks, a textbook is still the typical gateway to biology or chemistry or physics. As the sciences explode into subfields—making it less likely that any given professor will be expert in all the subjects she or he must teach—reliable textbooks become all the more important. In short, circumstances are forcing us to place ever-greater faith in science texts at a time when fewer and fewer seem to warrant it.

Epilogue: The Case of "Little Albert"

With the exception of Pavlov's dogs, no experiment has been cited more often in introductory psychology texts than James B. Watson and Rosalie Rayner's study of the emotional conditioning of an eleven-month-old infant, "Little Albert."[28] In their experiment, Watson and Rayner attempted to condition an emotional reaction to a specific stimulus and to see if that emotion would reappear in the presence of similar stimuli and persist through time.

Little Albert, an unusually "stolid and unemotional" baby, showed no fear when confronted with such stimuli as a white rat, a dog, a monkey, masks, and burning newspapers, but he did cry when a steel bar suspended behind him was suddenly struck with a hammer. Watson struck the bar whenever Albert touched a laboratory rat. After two trials on one day and five more a week later, Albert acted fearful in the presence of the rat alone. To see if that fear would occur also in the presence of similar stimuli, Albert was then shown a rabbit, a dog, cotton wool, human hair, a fur coat, and a Santa Claus mask. Five days later he was again shown the rabbit and the dog and, after waiting another month, the rat, the rabbit, the dog, a coat, and the mask. The timing, recording of events, and presentation of stimuli were, at best, haphazard.

In their original article, Watson and Rayner concluded that the experiments "show conclusively that directly conditioned emotional responses as well as those conditioned by transfer persist, although with a certain loss in the intensity of the reaction, for a longer period than one month."[29] Other investigators, however, were unable to replicate those results.[30] Moreover, when Albert's fear of the rat began to fade after ten days, Watson and Rayner decided to "freshen" the reaction by striking the bar while placing the rat on Albert's hand. Even so, his reactions in some later trials are more accurately described as ambivalent than fearful. Thumb-sucking, according to the original account, typically blocked the fear response. To obtain the predicted result, Watson and Rayner frequently had to remove Albert's thumb from his mouth.

Attempts to transfer the response also produced ambiguous results. Albert did not react negatively to all of the other stimuli. Moreover, when Watson and Rayner tried to condition Albert directly to the rabbit and dog, he showed only a slight reaction to both. The researchers themselves later conceded the tentative character of their results. In a footnote to another article they wrote: "The work at Hopkins was left in such an incomplete state that verified conclusions are not possible; hence this summary . . . must be looked upon merely as a preliminary exposition of possibilities rather than a catalogue of concrete usable results."[31]

Yet Little Albert soon became a model of the classical conditioning of emotional behavior. In the process, the story was also transformed—becoming in some ways simpler and in others more complex. Aspects of the study that might have cast doubt on its validity were omitted, while various imaginative details were introduced. Textbook accounts had Watson and Rayner humanely extinguishing the fears they created. Some authors even described that fictional process in detail; Albert is deconditioned by having the rat presented while being fed candy, or even more specifically "pieces of chocolate" or a "bowl of ice cream."[32] Both the details omitted and those added to the original result in a more favorable account of the experimenters and their study.

In the 1970s, the Little Albert story began to unravel. A number of articles noted discrepancies between the original article and textbook accounts.[33] The two most extensive exposés, published in the widely read *American Psychologist*, concluded that the Little Albert story was a piece of folklore. But the impact of these critiques was trivial. Of thirty texts that mentioned Little Albert, only one accurately summarized the criticisms,[34] and some continued to refer to stimuli specifically identified as fictitious, such as Albert's mother's coat and his teddy bear.

While authors did jettison stories about the extinction of Albert's fear, they often asserted that Watson and Rayner's intention to recondition the infant was thwarted by his removal from the hospital where the experiment was conducted. Indeed, the strong implication of some of these accounts is that the mother was to blame for removing her child in the middle of the study. One author claimed that, "Albert's mother . . . was so outraged by her son's fear of her (because of the [fur] collar) that she removed him before Watson could attempt to reverse the process."[35] However, Watson and Rayner knew a month in advance when Albert would leave and made no plans to remove the fears they themselves believed were "likely to persist indefinitely."[36] The fur collar is also fictional. Those authors who responded to the exposés at all typically corrected only the details that could be changed or omitted without undermining the story's function in the textbook. Most simply repeated the classic story, accompanied by the classic picture of Rayner with Watson in a Santa Claus mask.

As in the genetics case, critics are often liberally cited—but their points ignored. Why are they cited at all? Textbook authors are under considerable pressure to keep their references current. An author who cites older works will often be instructed by manuscript reviewers and editors to consult the current literature. Apparently, they sometimes respond simply by updating references. As a result, they may bear little relationship to the substantive discussions; indeed, they may directly contradict claims asserted in the text. Ironically, an important effect of updated citations is to provide new authority for discredited claims.

That publishers want to maximize profits while authors want to minimize time and effort helps explain the similarity of so many textbooks. But so does a utilitarian attitude toward the facts. Some authors have always been willing to tell stories they know are untrue when the moral seems fundamentally correct and the tale convenient for pedagogical purposes. But that practice is hardly restricted to textbooks.

Notes

1. Diane B. Paul, "Textbook Treatments of the Genetics of Intelligence," *Quarterly Review of Biology* 60 (1985): 317–26.

2. A particularly wonderful case, the persistence of the myth of the "Giffen Paradox" in economics textbooks, has been recently described in Terrence McDonough and Joseph Eisenhauer, "Sir Robert Giffen and the Great Potato Famine: A Discussion of the Role of a Legend in Neoclassical Economics," *Journal of Economic Issues* 29 (September 1995): 747–59. According to many texts, the British economist Sir Robert Giffen is said to have observed peasants consuming more potatoes as their price rose during the famine of the mid-nineteenth century. However, the paradox of an upward-sloping demand curve was debunked by George F. Stigler in 1947. When it nevertheless found its way into the 1964 edition of Paul Samuelson's famous introductory economics text, the legend took on new life. The story is widely recounted in texts notwithstanding the fact that it "is unsupported in the literature, implausible on the face of it, and more easily explained in other ways even if true" (p. 753).

3. Diane B. Paul and Arthur R. Blumenthal, "On the Trail of Little Albert," *Psychological Record* 39 (1989): 547–53.

4. Diane B. Paul, "The Market as Censor," *PS: Political Science and Politics* 21 (1988): 31–35; "The Nine Lives of Discredited Data," *The Sciences* 27 (1987): 26–30.

5. See Debra E. Blum, "Authors, Publishers Seek to Raise Quality and Status of College Textbooks, Long an Academic Stepchild," *Chronicle of Higher Education,* July 31, 1991, pp. A11–12; Diane Manuel, "Failing Grades for Textbooks," *Boston Globe,* March 1, 1992, pp. A1, 9–10.

6. John B. Jenkins, *Human Genetics* (Menlo Park, Calif.: Benjamin Cummings, 1983), p. 374.

7. H. Eldon Sutton, *Introduction to Human Genetics* (San Francisco: W. H. Freeman, 1980), p. 497.

8. Leon Kamin, *The Science and Politics of I.Q.* (Potomac, Md.: Lawrence Erlbaum, 1974).

9. Oliver Gillie, *Sunday Times* (London), October 24, 1979, p. 1.

10. The case against Burt was reopened by social scientists Robert Joynson in *The Burt Affair* (London: Routledge, 1989) and Ronald Fletcher in *Science, Ideology, and the Media: The Burt Scandal* (New Brunswick, N.J.: Transaction Publishers, 1991), both of whom assailed the motives of Burt's critics and argued that their charges of fraud were unproved. Like most reviewers of these books, I found the arguments tendentious and unconvincing. In any case, neither Joynson nor Fletcher argued that Burt's results were reliable.

11. L. Erlenmyer-Kimling and L. F. Jarvik, "Genetics and Intelligence: A Review," *Science* 142 (1963): 1477–79.

12. Quoted in Karen J. Winkler, "New Approaches Changing the Face of Textbook Publishing," *Chronicle of Higher Education,* May 16, 1977, p. 1.

13. Lewis A. Coser, Charles Kadushin, and Walter W. Powell, *Books: The Culture and Commerce of Publishing* (New York: Basic Books, 1982), p. 273.

14. Charles G. Morris, Preface to *Psychology: An Introduction* (Englewood Cliffs: Prentice-Hall, 1985).

15. Richard B. McKenzie, "The Emergence of Kickbacks in University Textbook Adoptions," unpublished manuscript, 1987.

16. Coser, Kadushin, and Powell, p. 210.

17. Robert Sommer, Marina Estabrook, and Karen Horobin, "Faculty Awareness of Textbook Prices," *Teaching of Psychology* 15 (1988): pp. 17–21.

18. Meredith Corporation v. Harper and Row, Publishers. Federal Supplement 378 (1974).

19. Coser, Kadushin, and Powell, p. 101.

20. Interview conducted by the author.

21. McGraw-Hill, Inc. v. Worth Publishers, Inc. Federal Supplement 335 (1971).

22. So has recent competition from instructional-software packages, packets of photocopied course readings, and desktop publishing.

23. Thomas W. Gornick, *Book Publishing Annual (1984)* (New York: R. R. Bowker, 1984).

24. Jenkins, *Human Genetics,* p. 375.

25. Robert H. Tamarin, *Principles of Genetics* (Boston: Willard Grant, 1982), pp. 291–92.

26. Anna Pai and Helen Marcus-Roberts, *Genetics: Its Concepts and Implications* (Englewood Cliffs: Prentice-Hall, 1981), p. 605.

27. Interview conducted by the author.

28. John B. Watson and Rosalie Rayner, "Conditioned Emotional Reactions," *Journal of Experimental Psychology* 3 (1920): 1–14.

29. *Ibid.*, p. 12.

30. Ben Harris, "Whatever Happened to Little Albert?" *American Psychologist* 34 (1979): 151–60; E. R. Hilgard and D. G. Marquis, *Conditioning and Learning* (New York: Appleton-Century, 1940).

31. John B. Watson and Rosalie Rayner Watson, "Studies in Infant Psychology," *Scientific Monthly* 13 (1921): 493–514.

32. Cited in David Cornwell and Sandy Hobbs, "The Strange Saga of Little Albert," *New Society,* March 18, 1976, p. 604; Robert E. Prytula, Gerald D. Oster, and Stephen F. Davis, "The 'Rat-Rabbit' Problem: What Did John B. Watson Really Do?" *Teaching of Psychology* 4 (1977): 45.

33. Cornwell and Hobbs, pp. 602–604; Prytula, Oster, and Davis, pp. 44–46; Harris, pp. 151–60; Franz Samelson, "J. B. Watson's Little Albert, Cyril Burt's Twins, and the Need for a Critical Science," *American Psychologist* 35 (1980): 619–25.

34. D. E. Papalia and S. W. Olds, *Psychology* (New York: McGraw-Hill, 1985).

35. S. A. Rathus, *Psychology* (1987), p. 219.

36. Watson and Rayner, p. 12.

4

THE ROCKEFELLER FOUNDATION
AND THE
ORIGINS OF BEHAVIOR GENETICS

Alan Gregg, director of the Rockefeller Foundation's Division of Medical Sciences, was sure that nature contributes more than nurture to differences in human behavior. Unfortunately, the public did not appear to share this conviction. Doctors and educators were particularly reluctant (or so Gregg thought) to admit the importance of genetics. In a 1945 letter to Clarence C. Little, then director of the Jackson Laboratory in Bar Harbor, Maine, Gregg expressed this opinion:

> In medicine, after years of attention to the nature of the invading organism and various aspects of advanced disease, we are beginning to realize that susceptibility to infection and tendency to degenerative diseases and various abnormalities of form and function have much to do with genetics. Since education is so extraordinarily important for all parts of society and all fields of endeavor, a more accurate definition of the limitations of education through planned experience of the

pupil would actually increase the effectiveness of education. There would not be such an incredible and insufferable amount of effort spent in present teaching methods wasted on material that cannot profit from such methods, and there might be a realization that man has as much to gain intellectually by wise matings as by $800,000 high schools.[1]

By 1945, Gregg had directed the Rockefeller Foundation's (RF) Division of Medical Sciences (MS) for fifteen years. He had first joined the foundation in 1919, as a young MD, and three years later was offered a position as Richard Pearce's assistant in the Division of Medical Education.[2] In 1928, the Rockefeller philanthropies entered a period of reappraisal and change. Raymond Fosdick, then the chief counsel to John D. Rockefeller, effected a consolidation of the four Rockefeller boards, in which all programs related to "the advance of human knowledge" transferred to the RF. The character and consequences of this merger have been described by others, in particular Robert Kohler.[3] Suffice it here to say that they signalled a shift from traditional Rockefeller concerns with education and applied social service to support for scientific research. This development was reflected in the new title of the Division of Medical Education, which in 1929 became the Division of Medical Sciences.

The review also prompted a change in the kind of research supported. In the past, funds had been awarded to the "best" individuals and institutions, with little regard to field. The new policy emphasized research that promised a substantial social return. In the view of Rockefeller trustees and officers, the times demanded planning, and planning, in turn, required a scientific understanding of behavior. However, a science of human behavior did not yet exist. The development of such a science would thus become the central focus of the Foundation. By 1933, the natural, social, and medical sciences divisions were officially joined in a project to develop "a new science of man," whose aim was the analysis and control of behavior.[4] A staff report of 1933 notes that all divisions should focus on two areas: the "conscious control of race and individual development with rather particular reference to mentality and temperament," and the "study and application of knowledge of social phenomena and social controls."[5]

These broad changes in the organization and character of the foundation were already reflected in the medical area by the mid-1920s. In 1925, RF trustees decided to devote some resources to the study of biology and psychology in relation to medicine and public health.[6] Within the Division of Medical Education, this policy was expressed in large grants to two German Institutes: Emil Kraepelin's biologically oriented Institute of Psychiatry in Munich and Oskar Vogt's Institute for Brain Research in Berlin.[7]

However, the pace of change greatly accelerated after Gregg succeeded Pearce as director of the MS Division on the latter's death in 1930. During Gregg's tenure, the Division drastically reduced its programs in medical education in order to support research in a new area variously titled "psychobiology," "mental hygiene," or later, "psychiatry." Under these rubrics, Gregg supported a wide variety of biologically oriented approaches—endocrinological, neurophysiological, and genetic—to the understanding of behavior.

This essay describes one strand in that program: work on the genetics of mental traits. RF-funded efforts in this area were instrumental in the development of the field that would come to be called "behavior genetics." Of course no single institution was responsible for the creation of this (or any) discipline. A number of wealthy amateur eugenicists, small foundations, and the American Eugenics Society (AES) all supported research on the genetics of behavior in the 1930s and 1940s. Their efforts, however, were dwarfed by those of the Rockefeller Foundation.

In the 1930s and 1940s, the government provided little support for research, apart from that conducted at its own institutions. Thus, the private foundations contributed most of the funds available for research by non-government scientists. Among them, the Rockefeller Foundation was paramount. In the 1930s, it accounted for more than one-third of foundation giving to all fields, and nearly three-quarters of funds expended in support of research in the natural sciences.[8]

The decision to promote work on the genetics of behavior reflected certain social commitments of RF trustees and officers. By the early 1930s, when the "science of man" program was established, the eugenics movement had already come under attack within the scientific community. Many of those who wished to distance themselves from the movement's scientific crudity and reactionary politics, however, shared its strong hereditarian assumptions and at least long-term commitments to the breeding of a better race. These people not only associated eugenics with scientific naïveté and open propaganda; they tended to *define* eugenics as a movement tainted by these failings. Up-to-date research, whatever its aims, thus could not be eugenics. And the officers of reorganized RF always saw themselves on the cutting edge of science.

Behavior genetics, even in the period 1930–1950, was not merely the old eugenics movement under a new name. There was, however, a strong underlying continuity of beliefs and commitments. The field of behavior genetics emerged from the efforts of various institutions—most importantly the RF—to demonstrate the falsity of environmentalist assumptions. The character of this quest (and its uncertain success) will be illustrated with a case study of the first major American behavior genetic project: "Genetics and the Social Behavior of Mammals," begun in 1945 at the Jackson Laboratory in Bar Harbor, Maine. By the time Rockefeller funding ended, eleven years later, this study had absorbed more funds than any other genetics project supported by either the MS or NS Division. The "science of man" program, with its emphasis on behavior, began in 1933. Why did the RF wait until 1945 to support American research in behavior genetics? The answer to that question lies in the complicated history of earlier Rockefeller efforts in human biology.

Origins of the Rockefeller Program

It is sometimes said that the RF of the 1930s spurned applied human biology, which was viewed as a throwback to the older emphasis on the application of science to social service and reform and hopelessly entangled with eugenics.[9] That RF officers wished to have nothing to do with eugenics is especially stressed in Gerald Jonas's recent book on the Foundation's role in the development of modern science. According to Jonas, Max Mason, who became president of the reorganized Foundation in 1928, determined from the start not to support eugenics. His rejection of that program was "unequivocal."[10]

It is true that, even before the reorganization, attitudes of RF officers toward eugenics were mixed. Thus, in the 1920s, Edwin

Embree, director of the old RF's Division of Studies, was thwarted in his efforts to develop a eugenically oriented program in human biology. In 1925 Embree told Raymond Fosdick, the Rockefeller family representative on the RF Board of Trustees and a member of the advisory committee of the AES, that he was:

> tremendously interested in the sciences of human biology, the possibilities of which we are beginning to explore. If it is possible to do anything in such matters as eugenics and a better understanding of mental processes, we shall be making contributions indeed. I realize that it is going to be much more difficult to take bad ideas out of people than it has been to extract hookworms; harder to give them good inheritance than good surgery. While a more complicated undertaking, it is also even more worth investment and speculation.[11]

Embree's plans were opposed, however, by Richard Pearce, the director of the Division of Medical Education, and Simon Flexner, head of the Rockefeller Institute, both of whom apparently believed that the prestige of academic biology and genetics was rising while that of eugenics and mental hygiene was in decline.[12] Pearce and Flexner succeeded in convincing Fosdick not to support a large-scale project in this area. Although he lacked trustee endorsement for his efforts, Embree did not give up. On an extended trip to Europe in 1926, ostensibly in connection with nursing education, he met with biologists to promote his program. Frustrated by resistance at the Foundation, however, he soon resigned.[13]

With the appointment of Mason in 1928, eugenics—at least of the mainstream kind—fell still further from favor. In 1929, Mason rejected a proposal from the Eugenics Research Association for a project on human inheritance, race mixing, and the differential birth rate.[14] His protégé, Warren Weaver, who was appointed director of the Natural Sciences Division in 1933, adopted an equally critical stance. As a result, the RF declined both Frank Lillie's and C. C. Little's proposals for support of institutes for the study of biology and social problems. But the refusals do not constitute the whole story of the RF and eugenics. As we will see, proposals that were explicitly eugenic *were* funded in the 1930s, although not in Weaver's division and not in the United States. Moreover, analysis of the rejected proposals indicates that even

Weaver's attitude cannot be characterized as a simple rejection of eugenics.

In 1931, Lillie wrote to Mason, resurrecting a proposal for an "Institute of Racial Biology" that he had first suggested in 1924. Indeed, he merely enclosed the original memorandum with his letter to Mason. According to Lillie:

> [T]he future of human society depends on the preservation of individual health and its extension into the field of public health; but it depends no less on social health, that is the biological composition of the population. We are at a turning point in the history of human society—the age of dispersion and differentiation of races is past. The era of universal contact and amalgamation has come. Moreover, the populations press on their borders everywhere, and also, unfortunately, the best stock biologically is not everywhere the most rapidly breeding stock. The political and social problems involved are fundamentally problems of genetic biology.[15]

Weaver considered Lillie's suggestion compatible with RF aims and deliberated the question of funding for several years. Since it would promote basic research, neither Weaver nor Lillie associated the proposed institute with eugenics. Indeed, Lillie wrote that "it should be kept free of all propaganda concerning eugenics, birth control, etc.; and in such connections aim merely to furnish the indispensable scientific foundations on which social prophylaxis of the future must depend."[16]

A number of factors contributed to the project's ultimate rejection. It violated Foundation policies both against committing funds for long periods and endowing large institutions.[17] The institute would be devoted to a specific problem—an approach rejected by Raymond Fosdick.[18] The year 1931 was also a very poor time for establishing enterprises that required a substantial infusion of funds.

Six years later, Little advanced a somewhat similar proposal. In 1937, he met with the directors of the three RF divisions (Edmund Day of the Social Sciences Division, Weaver, and Gregg) to discuss the possibility of Rockefeller funding of an Institute of Social Biology and Medicine. The follow-up proposal reflects an array of loosely linked concerns: genetics in medicine, the physiology of sex and contraception, human psychology, growth and development, "population problems," eugenics. The prose is dramatic (there is a "terrifyingly urgent . . . need to preserve the sex cells of civilization—the centers of creative initiative before our overfed and undernourished civilization becomes a great uninspired eunuch with no power to generate the units that build the future"), but neither the content nor the relationship of these concerns is further defined.[19] Thus, the failure of these two programs imply little about the RF stance toward the subject of eugenics as such.

The severest anti-eugenic remarks were voiced by Warren Weaver. For example, in a 1933 report he wrote that "work in human genetics should receive special consideration as rapidly as sound possibilities present themselves. The attack planned, however, is a basic and long-range one, and such a subject as eugenics, for example, would not be given support."[20] He also asked (rhetorically) "whether we can develop so sound and extensive a genetics that we can hope to breed, in the future, superior men?"[21] These comments are not necessarily contradictory. Weaver believed that "the human race needs, and needs desperately," a science of human genetics, which would ultimately be used to produce a better race.[22] His contribution to this end, however, would be the funding of basic work in mammalian genetics.

Galton's own definition of eugenics—"the study of agencies under social control which may improve or impair the racial qualities of future generations"—would seem to describe Weaver's efforts. Weaver himself obviously applied a narrower definition. But this is not the place to consider how Weaver's activities are best characterized, for Weaver was not responsible for funding work in human genetics. That field was the province of Alan Gregg, whose Medical Sciences Division was responsible for research with human subjects. Gregg has received short shrift in the growing literature on the RF. Scholars have been far more interested in the articulate (and self-promoting) Weaver.[23] In the story that concerns us, however, Gregg is the crucial actor.

Alan Gregg, Psychobiology, and Human Genetics

Gregg's program in "psychobiology" (later called "psychiatry") absorbed about three-fourths of the funds expended by the MS division.[24] In Gregg's view, the costs of our failure to understand the workings of the human mind were manifest in the "economic, moral, social, and spiritual losses occasioned by the feebleminded, the delinquents, the criminal insane, the emotionally unstable, the psychopathic personalities," as well as in the less extreme but far more common (and preventable) fears, phobias, and aberrant behavior of otherwise sane human beings.[25] In many countries, he argued, more beds were devoted to the care of mental cases than to all other diseases combined. He also considered the educational system to be enormously wasteful. (Indeed, Gregg's correspondence and internal memoranda actually focus much more on the failures of education than on medicine.)

The new field of psychobiology, designed to address these problems, encompassed various approaches to the understanding of mind. Work in human genetics constituted only one component in a multifaceted research program. But given Gregg's belief that differences in human cognitive abilities as well as susceptibility to mental illness were largely attributable to differences in genes, and his assumptions about the relevance of such differences for social policy, it was an important element.

In the 1930s, however, there existed few attractive opportunities in the United States to promote research on the genetics of mental traits. American psychiatrists had little interest in biology, and virtually none in genetics. Geneticists, on the other hand, focused on corn and fruit flies, and ignored humans. At least that is what RF officers believed. Thus, in a 1936 report that includes an extensive assessment of the state of American genetics, Charles B. Davenport of Cold Spring Harbor is characterized as the "leading American worker in human genetics."[26] He is also the only American mentioned. In the 1930s, therefore, Gregg looked to Europe.

He considered the Scandinavian countries particularly suitable for work in human genetics given their homogeneous and stable populations and the existence of accurate and complete medical records. Beginning in 1930, a number of small grants in human genetics/eugenics were approved by the RF Paris Office to the Pathological Institute of Copenhagen, directed by Oluf Thomsen. His student, Tage Kemp, also received two RF fellowships (one enabling him to work with Davenport at Cold Spring Harbor) and a grant in 1934 in support of work on the genetics of psychopathology. In 1936, Gregg appropriated $90,000 toward the establishment of an Institute of Human Genetics, directed by Kemp, at the University of Copenhagen. This institute was to engage both in research, especially on the heritability of mental traits, and in genetic counseling explicitly informed by eugenic concerns.

Germany was thought to be particularly advanced in research on the area of genetics of mental traits. As a result, it received a large share of the funds expended by the MS Division. Between 1930 and 1935, Gregg contributed $125,000 for a nationwide anthropological survey of the German people directed by Eugen Fischer. The project was undertaken, according to an RF report, in order "to provide a means of finding a scientific basis for the study of the racial or biological composition of the German people and of the interaction of biological and social factors in determining the character of the present population."[27] Between 1932 and 1935, the RF also appropriated funds for twin research and for studies of the effects of poisons on the germ plasm at Fischer's Kaiser Wilhelm Institute (KWI) for Anthropology, Human Genetics, and Eugenics. Other genetic/eugenic studies at the KWI for Brain Research and the German Psychiatric Institute (under Ernst Rudin, an author of the German sterilization law) also continued to receive RF funds even after Hitler's seizure of power. In 1939, the foundation finally ended support of all programs in Germany. With the exception of two grants to Viennese social scientists, the only projects still being funded were in Gregg's psychiatry program.[28]

Foundation funds in Britain were channeled through the Medical Research Council, which in the mid-1930s elaborated a new program in psychiatry and related subjects to be directed by a Mental Disorders Committee. During the 1930s, the MS Division supported projects by D. K. Henderson and T. A. Munro at Edinburgh on consanguineous marriage and mental disorders, Janet Vaughan at Hammersmith on human heredity in relation to psychic disturbances and neurological diseases, Lionel Penrose for the Colchester Survey on mental deficiency, and R. A. Fisher for serological research.

Work in serological genetics was considered especially exciting. In the early 1930s, it seemed likely that serological analysis could be used to identify heterozygote carriers of genes responsible for mental defect. If heritable antigens in the blood were linked to such genes, they could serve as genetic "markers" for the traits of interest. The identification of such markers would solve a problem that had bedevilled eugenics since the 1910s: the fact that most of the genes responsible for mental defect were hidden in apparently normal carriers and thus policies to prevent only the affected themselves from breeding would work very slowly.

In the 1910s and 1920s, some eugenicists hoped to overcome this obstacle by making use of the fact that many "recessive" genes in fact have slight phenotypic effects, i.e., the phenomenon of partial dominance. If a normal mentality were not completely dominant over a defective one, heterozygote carriers might yet be identifiable, for example through mental tests. In the late 1920s, however, the invention of serological genetics appeared to promise a far more straightforward approach to this problem than one relying on human symptoms. Some geneticists hoped that those recessive genes causing mental defects when homozygous would themselves have serological effect, and thus be directly detectable.[29] Even in the absence of any effects specific to these genes, however, it was assumed that they would be closely linked with others occurring on the same chromosomes. Given the rapidly increasing number of recognized serological reactions, the prospect of finding one or more of the genes linked to those responsible for a given defect seemed quite promising—provided funds were available for carrying out systematic research in pedigrees exhibiting anomalies. Thus, Gregg supported a number of projects in serological genetics, in Denmark and Germany, as well as Britain.[30]

Turning to America

During the war, Gregg hoped that the Carnegie Institution would eventually become a major source of support for work in human genetics. After 1917, the Carnegie had taken over funding of the Eugenics Record Office (ERO), then the leading eugenics institute in the United States. By the mid-1930s, however, the Carnegie trustees had become disenchanted. Among other embarrassments, the ERO's superintendent, Harry Laughlin, insisted on praising Nazi eugenic policies. In 1939, the ERO was closed.[31] Gregg thought that Frederick Osborn, a Carnegie trustee who was also the secretary of the American Eugenics Society, might persuade it to reenter the field. Osborn had been called to Washington during World War II, however, in order to direct the Army's division of education and propaganda. In 1944, it looked as though Osborn might stay in Washington for some time, making it unlikely that the Carnegie would become a major force in human genetics. In Gregg's view, the RF needed to take up the slack.[32]

In 1945, he awarded the first in a series of substantial grants to the psychiatrist Franz Kallmann for research on the genetics of schizophrenia. As a half-Jew, Kallmann had been removed from his position at the KWI for Biology in Berlin in 1935. The following year he emigrated to the United States and found a position at the New York State Psychiatric Institution. His research there was supported by the Carnegie Foundation.

Kallmann's views on the "nature-nurture" question were extreme even for the 1930s. In 1935, while still in Germany, he proposed to extend the compulsory sterilization law of 1933 to include the heterozygous carriers of the gene for schizophrenia. Kallmann believed that these apparently healthy carriers would exhibit minor anomalies and thus be detectable. He therefore proposed testing all the close relatives of schizophrenics. The testing program was to have been so massive, and would have involved the consequent sterilization of so many people, that it was considered impractical even by the Nazis.[33]

In the same year, Gregg also made a large grant to the Bar Harbor project. Its goal—or rather Gregg's—was conclusively to demonstrate a high heritability of human intelligence. The subjects in this study were dogs.

In a contemporary perspective, the use of dogs to substantiate claims about human behavior may appear somewhat peculiar. As we will see, however, in 1945 there was no consensus that humans make the best subjects for human genetics, even the genetics of mental traits. And mental traits themselves were a central concern of work in medical genetics during the inter- and immediate postwar periods. There was, in fact, no real distinction between medical and what would later be called behavior genetics. The emphasis on mentality, in turn, reflects the field's eugenic origins.

Medicine and Mentality

At least in the Anglo-American world, eugenicists had always emphasized the importance of mental, rather than physical characteristics. Most eugenicists were convinced of the heritability of virtually all aspects of intellect, personality, and character. Few would demur from the view expressed by the author of an early genetics text that "musical, literary, or artistic ability, for example, mathematical aptitude and inventive genius, as well as cheerful disposition or a strong moral sense are probably all gifts that come in the germ plasm."[34]

Of all these traits, intelligence was generally the most highly valued, and its purported decline the cause of the greatest concern. In the United States, the primary source of degeneration was thought to be new immigrants from southern and eastern Europe. The American eugenic literature of the 1910s and 1920s abounds in comparisons of the reproductive rate of Harvard students with that of immigrant groups.[35] At the center of the debate over immigration restriction (to the extent that it concerned biology) was the problem of "feeblemindedness." Of course eugenicists were concerned with other traits as well, such as criminality and shiftlessness. But they emphasized mental defect, thinking it the root cause of most other social problems. Some eugenicists believed that there were special genes for criminality. Most, however, assumed that people became criminals because they were feebleminded. In contrast, questions of health and disease received short shrift in the eugenics literature.

The emphasis on mentality carried over to post–World War II efforts to develop a science of medical, or more broadly, human genetics. In the decade following the end of the war, this field had relatively less to do with studies of clinical disease than of intelligence and temperament. That is not surprising when we realize that most of the pioneers in the field of human genetics—and their patrons—were active eugenicists. Five early presidents of the American Society of Human Genetics (which was founded in 1948) served on the Board of Directors of the American Eugenics Society (AES). Indeed, the AES itself played an important role in promoting and subsidizing both research and publications in the field. Its focus, however, was on intelligence and personality, not disease. The Society's director, Frederick Osborn, declared in 1948 that he "would not emphasize physical health as a direct objective of the eugenics effort." In Osborn's view, physical factors would take care of themselves. If those incapable of "sound thinking" could be induced not to breed, "that would take care of their physical characteristics at the same time. . . . Our practical program of eugenics needs then to be concerned only with mental qualities."[36]

A 1954 editorial in the Society's journal likewise asserted that negative eugenics, aimed at reducing disease, is of far less importance than a positive program to raise the level of the population in respect to mental traits.

The great possibilities for improvement lie in changing the distribution of births among normal people, so as to increase the proportion of children at the higher levels of normal capacity, while reducing the proportion of those at the lower levels of normality. This would raise the average level of the whole and greatly increase the possibilities for a better human society and for individual and social happiness. This is the great field for eugenic advance, and here lies the opportunity of the Eugenics Society.[37]

It should be noted that Osborn headed not only the AES but also the Rockefeller-funded Population Council, which supported most American twin research in the 1940s and 1950s.

Thus, in the perspective both of the scientists and of those who funded their work, medical genetics had a very broad meaning. The genetics of mental deficiency, normal intelligence, and schizophrenia received at least as much attention as the genetics of clinical disease. This concern with intellectual and emotional variation was justified on two grounds. On the one hand, differences in susceptibility to disease were thought to have a large emotional component, which was itself highly heritable. On the other, work in human genetics was expected to inform genetic counseling. For counseling to be effective, it was essential to have a genetic picture of the individual as a whole. Most counselors believed it unwise to advise against reproduction if the individual possessing the defective gene was above average in character and intelligence. It was the "total genetic potential" that mattered, and not just the obvious abnormality.[38] Thus, in the immediate postwar as in the inter-war period, medical genetics *was,* in large degree, behavior genetics. A study of intellectual and emotional variation was therefore perfectly congruent with medical genetics, as it was conceived in the 1940s.

But intelligence and temperament in dogs? In late 1945, after the project was already underway, Gregg talked with R. A. Fisher in London. According to Gregg's diary, Fisher "said that one of the most valuable things that could be done in genetics would be work with dogs in studying temperament and nervous disposition."[39] The assumption that one could (and should) generalize from the behavior of other organisms to humans would have appeared plausible to many of Gregg's contemporaries.

In the 1940s, some of the chief contributors to the field of "human genetics" worked with non-human organisms. Thus, Curt Stern, author of the influential textbook *Human Genetics,* worked with *Drosophila.* So did Hermann J. Muller, the first president of the American Society of Human Genetics, while Hans Nachtsheim, the most prominent German in this field during the immediate post-war period, worked with rabbits. The indirect approach to human genetics was strenuously defended by A. H. Sturtevant in his 1954 Presidential Address to the Pacific Division of the AAAS. According to Sturtevant, "man is one of the most unsatisfactory of all organisms for genetic study." He argued that:

There are enough unambiguous examples known to make it clear that the same principles are at work in man as in all other higher animals and plants—and even without such evidence, enough is known about the cytology of human tissues to give us confidence that no peculiar kind of inheritance is to be expected in man. In fact, much of the argument concerning the practical aspects of the genetics of man is best based on experimental evidence from other organisms rather than on what is known directly from study of human populations. . . . The position is especially unsatisfactory with respect to the most important of all human differences—namely, mental ones.[40]

There was, moreover, nothing new about the use of dogs in studies of human biology. Indeed the RF had already funded an earlier genetics project with dogs: Charles Stockard's experiments testing whether breed differences in dog anatomy were attributable to hereditary disorders in the ductless glands. Nor was the Bar Harbor project the first to utilize dogs to study human mental traits. The most famous example—well known to foundation officers—was Ivan Pavlov, who in 1929 began to relate his work on conditioned reflexes and experimental neuroses in dogs to human mental disease.[41] For about six years during the 1920s, the American psychologist W. Horsley Gantt worked with Pavlov at his Institute of Experimental Medicine. Like Pavlov, Gantt wished to use dogs to say something about human psychology. Unlike Pavlov, he was much concerned with uncovering the genetic basis for differences in the dogs' temperaments. One of the visitors to Pavlov's laboratory was Alan Gregg. Gregg was greatly impressed with Gantt's work and recommended him for a staff position at the Johns Hopkins Medical School, where Gantt did move in 1929.[42] From 1931 to the mid-1940s, the MS Division provided most of the funds for his research.[43]

Dogs seemed particularly appropriate for Gregg's postwar project. As noted earlier, he was convinced that a rational social policy depended on a clear and compelling proof of the falsehood of contemporary environmentalist assumptions. Educators, doctors, and the general public had all to be convinced. The problem, as Gregg came to see it, was that the heritability of behavioral traits had been demonstrated in organisms—such as fruit flies and rats—to which few persons could relate emotionally. The solution lay in making the point with an animal that exerted a strong emotional appeal. From this perspective, dogs were ideal. Gregg was thus led to approach the geneticist Little, a former Harvard classmate and son of a dog fancier, whose cancer research with mice was already supported by the foundation.

C. C. Little and the Founding of the Jackson Laboratory

In 1929, Little resigned under pressure from the presidency of the University of Michigan. He had espoused birth control, tried to reform the university without the support of the faculty, and obtained a divorce. None of these were popular activities. Reporting on the reasons for his departure, the school newspaper noted that, "As a biologist, Dr. Little has been a strong advocate of race betterment programs and the science of eugenics and his courageous statements of his views on controversial subjects, such as scientific consideration of the question of birth control, have led to criticism in some quarters."[44]

Fortunately, Little had wealthy friends in the Detroit business and industrial establishments. Prior to his stint at Michigan, he had been president of the University of Maine. The university ran a summer course on Mt. Desert Island, where Edsel Ford, Roscoe B. Jackson (president of the Hudson Motor Company), and George Dorr (one of the Island's largest land owners) were summer residents.

Jackson, his brother-in-law Richard Webber (who owned the J. L. Hudson Department stores in Michigan), and Ford offered to finance a private institute for cancer research on land provided by Dorr. Although Jackson soon left for Europe, where he contracted typhoid fever and died without a will, the project was continued by his widow and the others. With the Depression, these private funds were substantially reduced and Little turned to the RF for support.

During its first decade, the Jackson Laboratory was devoted exclusively to cancer research; the RF provided funds for building, research, and maintenance of a mammalian stock center. However, Little's eugenic concerns remained strong and would soon intersect with an interest of Gregg's.

In late 1941, Gregg received a letter from Little proposing that the Laboratory breed a uniform strain of dog to be used in cancer-related experiments. Gregg's response was favorable, but he noted that a breeding program might also serve purposes that were, in all likelihood, incompatible with the project suggested by Little. Were it possible to test intelligence in dogs, Gregg reasoned, it should also be possible to breed a pet that was both amiable and extremely smart. In his view, anatomy had been stressed to a ridiculous degree. People actually appreciated intelligence and good disposition far more than anatomical features in dogs. Gregg was sure that there would be a substantial market for a dog that was both friendly and bright. He also believed that there would be two valuable byproducts of a project to breed such a pet: (1) their owners would be led to see the relevance of genetics to intellectual performance and, (2) they would come to feel indebted to experimental biology for having produced such a pet rather than be critical of the cruelty of animal experimentation.[45]

In 1941, Gregg was merely speculating on the possibility of testing and breeding for intelligence in dogs. But the idea continued to simmer, and two years later he wrote Edwin B. Wilson of the Harvard School of Public Health asking whether he thought that methods used with rats to measure and breed for intelligence would also work with dogs. Wilson liked the idea and suggested talking it over with Little, whom he had seen acting as a judge in the Boston Dog Show. Gregg was delighted with Wilson's reply. He was particularly pleased that Wilson had grasped his point about the value of demonstrating the heritability of intelligence in dogs. In Gregg's view, most people would never be impressed with the demonstration that there are more intelligent and less intelligent rats. They would simply "dispense with that phenomenon in much the same way as we do with trained fleas."[46]

Soon after, Gregg wrote to Little. He noted that while rat work was useful, it was also inaccessible to ordinary people. These people would, however, recognize and appreciate an intelligent dog. Current breeders refused to produce such a pet, fearing that their "morphologically perfect animals" would no longer be in demand. Gregg asked whether it would not be worthwhile to spend fifteen or twenty years breeding an extremely smart but small dog, "just to show that genetically intelligence is capturable and reproducible."[47] In this letter, the social agenda is bluntly described. "My point of departure," he wrote,

is a conviction that one of the constant afflictions of educators is their ignorance of the hereditary equipment of their pupils. Educators think that environment is everything, but it is not. Consequently, a great deal of their effort is wasted or worse . . . I'd like to see the talents of some geneticists devoted to the task of showing in a clear and readily accessible form that there are some aspects of intelligence which are . . . transmitted hereditarily . . . if as a result of some such effort a highly intelligent, scientifically tested animal were to be available for any considerable number of Americans and if such an animal was conspicuously intelligent and satisfactory as a pet, I believe the inference would be almost inescapable that in human beings also intelligence is affected by heredity and that the limitations of education in certain instances are clearly coming from genetic rather than pedagogic sources.[48]

Little's reply was enthusiastic. In his view, the dog would make an ideal experimental subject. Fanciers had already established striking and important differences among breeds. Moreover, dogs are much easier to breed, and much more prolific, than are any of the primates. He affirmed the social value of the proposed project as well. "If we are not convinced of the importance of individual variation and if we do not understand how it arises and how to utilize it," he wrote, "we shall never be able to create a democracy that will have in its own make-up the characteristics necessary to criticize it and to shape its destiny as it evolves."[49] Robert Yerkes also endorsed the use of dogs, noting, among other factors, that they make a strong emotional appeal to humans without arousing religious prejudice or superstitious bias.[50]

Gregg was, in any case, now prepared to move ahead. He informally promised Little $50,000 a year for ten years, plus another $50,000 to set up a laboratory. (The RF ultimately awarded $632,000 to the project, over a total of eleven years.) He also asked him to recommend a scientific director.

John Paul Scott

Little had been a student of William E. Castle's at Harvard. So had Sewall Wright, at about the same time, and the men were well acquainted. Two of Wright's former Ph.D. students at Chicago— Elizabeth Shull Russell and William L. Russell—were already members of the small staff of the Jackson Laboratory. It was thus natural for Little to turn to Wright for advice concerning an appropriate director. Wright recommended John Paul Scott, a thirty-five-year-old former Chicago student of his, with a strong interest in the genetics of behavior. Indeed, Scott was at the time the only American with a Ph.D. in genetics who was interested in the behavior of mammals. At the invitation of the Russells, he had already worked two summers at the Laboratory on a study of the differences in fighting behavior of among males in different inbred strains of mice.[51]

At the age of fifteen, Scott had read Albert Wiggam's popular eugenic tract, *The Fruit of the Family Tree*. In an autobiographical essay, Scott characterized the book as "old-fashioned eugenics, based on the simple theory that all the world's ills were due to bad heredity, and if we would only apply our knowledge of animal breeding to humans, Utopia would follow." He also wrote that, "I did not really swallow this, naïve as I was, but it did occur to me that if heredity had all that an important effect on behavior, someone ought to study it scientifically. And so began my interest in behavior genetics."[52] It is thus not surprising that, while working on his Ph.D. at Chicago, Scott should have been influenced by the ecologist/animal behaviorist W. C. Allee. He even did a bit of behavioral genetics in Allee's laboratory, with *Drosophila* stocks borrowed from Wright. That work caught the attention of Theodosius Dobzhansky, who suggested a followup study using a more sophisticated technique. Like Dobzhansky, Scott was interested in the genetics of other organisms primarily for what they could teach about humans. Unlike Dobzhansky, Scott was convinced that work of human relevance would have to be done with mammals.[53]

After obtaining his degree in 1935, Scott accepted a position at Wabash College in Indiana. There he became convinced that the social sciences were unscientific because they ignored biology. More than that: the solution to major social problems lay in the application of biological concepts to social phenomena. Scott reports that as the idea took hold, he "began to feel a little like the apostle Paul on the road to Damascus."[54] He determined to begin work on a new interdisciplinary science, debating whether to call it biosociology or sociobiology (finally settling on the latter). He understood, however, that he would not be taken seriously by social scientists unless he could "speak their language."[55] Thus, Scott and his wife moved to Boston for a year, where he studied in various libraries and began to write a book on social organization in humans and other animals. In 1939, he returned to Wabash, where he did both field and laboratory research on animal behavior and continued to write his book (without finding a publisher) on the relevance of animal behavior to human affairs.

That was the situation in 1945, when he was invited by Alan Gregg to direct a study of the genetics of behavior in dogs. The Jackson Laboratory was a weak and struggling institution. There was no tenure for the staff, or firm support. But Scott was enthusiastic about research with dogs, both because of the enormous variability among and within breeds, and because dogs are "timid or confident, peaceful or aggressive, and may be born with undershot jaws, club feet, or hemophilia"; that is, they vary in just the same respects as do humans.[56] Allee advised him to accept, and Little promised him a free hand to set up the research project and choose his associates. Scott agreed to go and invented a title for himself: chairman of the Division of Behavior Studies.[57]

The Project

In the summer of 1946, the RF funded a conference on "Genetics and Social Behavior" at the Jackson Laboratory. Two RF officers,

Alan Gregg and Robert Morrison, were in attendance. So were scientists from a variety of fields; indeed, they constituted a veritable Who's Who of research on behavior. Social psychology was represented by Gardner and Lois Murphy; comparative psychology, by Robert M. Yerkes (who also chaired the conference), Theodore Schneirla, O. H. Mowrer, and C. R. Carpenter; physiology, by Frank Beach and Benson Ginsburg (another Wright student, who had been an assistant of Allee's at Chicago and would later work on the project as a summer investigator); and genetics, by Clyde Keeler, C. S. Hall, H. H. Strandskov, and John L. Fuller. Scott joined the project full-time the following year; he would later co-author the first text in the field of behavior genetics.[58] Science journalism was represented by Walter Kaempffert of the *New York Times* and Gobind Lal of the Hearst papers, among others.[59] The scientific conferees were asked for advice on how best to proceed with the project; the journalists, how best to publicize the results. The latter also constituted a "Committee on Social Interpretation," whose job was to ensure that the public not only heard about the study but understood its point.

For thirteen years, Scott and Fuller collected data on five breeds of dogs: African basenjis, beagles, American cocker spaniels, Shetland sheepdogs, and wire-haired fox terriers. (Purebreds with very different behaviors but similar size were chosen in order to use the same apparatus for all). In the first phase of research, the five breeds were raised in the same environment, and their similarities and differences measured. At the same time, to study the development of behavior, daily observations of the puppies were recorded from their birth to sixteen weeks of age. In the second phase, the basenjis and cocker spaniels were crossbred, using a classical Mendelian design.[60] In all, some 300 puppies were rated in thirty major tests (each of which included multiple measurements); factor analysis and analysis of variance were applied to at least 8,000 separate pieces of data. The results of these analyses were unexpected, both to the investigators and the RF.

In the first place, correlations among the different behavioral tests were low. Breed differences existed, but the same breed would do very well in one test and badly in another; none was distinctly better in problem solving. Thus, no breed could be said to be smarter than another. An individual from any breed was able to perform well in a situation where it could be motivated and for which it had the physical capabilities. Within breeds, the capacities of individual animals were highly variable, but most animals could perform the required tasks by mobilizing different capacities.[61] The experimenters thus concluded that "nothing like the general-intelligence factor sometimes postulated for humans" exists for dogs.[62]

Nor did they find evidence of general temperamental factors. Breeds fearful in one situation were confident in another. Breed differences were also strongly influenced by habit and training. For example, raising puppies of two breeds together reduced their differences. In general, the behavior of young animals was highly variable within individual animals and strikingly similar among them. Thus, it appeared that genetic differences do not appear early in development, to be modified by later experience, "but are themselves developed under the influence of environmental factors and may appear in full flower only later in life."[63]

The responses to problem-solving tests were greatly affected by emotional and motivational factors. It was extremely difficult to separate these reactions from a tendency to perform well in a particular situation. On the basis of their heritability estimates for various traits in dogs, Scott concluded that the human estimates, especially for intelligence, "appear to be far too high." Indeed, he even speculated that "human differences in 'intelligence' reflect only differences in motivation rather than cognitive capacities."[64] What could the RF do with such results?

In a 1944 memo of his interview with Little, Gregg wrote of their agreement on the project's objective: the clear demonstration "to large numbers of persons of the fact that intelligence and other valuable qualities are not created by education as much as brought out by it and that the effectiveness of educational measures is definitely qualified by the inherent potentialities of the recipients thereof."[65] In correspondence, internal memoranda, motions to appropriate funds, and even the published RF annual reports, the primary objective is always defined as a demonstration of the limits of mass education.[66]

Scott and Fuller did not provide such a demonstration. Even worse, they called constant attention to what *they* saw as the social implications of the study's results. In 1956, at the project's conclusion, these were summarized by the investigators as follows: "The behavior traits do not appear to be preorganized by heredity. Rather a dog inherits a number of abilities which can be organized in different ways to meet different situations. . . . This means, in terms of human behavior, that the best sort of social environment is one which permits a large degree of individual freedom of behavior. Most individuals can reach desired goals if they are allowed sufficient freedom in the way in which they reach these goals."[67] Given these conclusions, it will perhaps come as no surprise to learn that the "Committee on Social Interpretation" was never mobilized.

Conclusion

In this essay, the RF appears as a kind of helpless giant. Notwithstanding Gregg's clear social agenda and near total control of the purse strings, he was unable to obtain results useful for his purposes. Thus, even before the development of federal funding and the advent of peer review, scientists seem to have achieved considerable autonomy. This conclusion needs to be qualified, however. Had the results of the project been more to Gregg's liking, there existed a committee to ensure that its message was brought to the public. As it turned out, the study was effectively buried.

The trend of behavior genetics generally has been much more compatible with Gregg's than with Scott's assumptions. The direction of the field may be explained, at least in part, by the social agenda that informed the decisions of the RF and other financial patrons in the field's early years. Most of the original investigators were funded by some combination of the RF, the AES, or smaller, right-wing foundations such as the McGregor and Pioneer Funds (as well as by private patrons, usually wealthy amateur eugenicists). The officers of these organizations were often linked by close personal relationships as well as by overlapping institutional memberships. Thus, Frederick Osborn, the director of the American Eugenics Society, was also a trustee of the Carnegie Corporation and president of the Population Council, which was founded in 1952 and fully funded by the Rockefeller family. There were, of course, differences in outlook and emphasis among these sponsors. But all could agree with the authors of a Population Council report that "social science research of the post-war period has been disproportionately devoted to studies of the effects of differences in the environment with no regard to differences in the genetic material on which the environment act."[68] And they all sought to remedy this "imbalance."

The funding process itself, as we have seen in the case of the Jackson Laboratory, was highly personalized. When mistakes were made, they were thus difficult to reverse. However, patrons were usually able to identify scientists whose views were congruent with their own. The strong hereditarian thrust of mammalian behavior genetics should thus be no surprise.

Notes

Citations to documents from the records of the Rockefeller Foundation, Rockefeller Archive Center, North Tarrytown, New York, provide Record Group, Series, Box, and Folder numbers.

1. Alan Gregg to C. C. Little, March 16, 1945; RF 1.2, 200A, 134, 1190.

2. A more detailed discussion of Gregg's background and career can be found in Wilder Penfield's breathless biography *The Difficult Art of Giving: The Epic of Alan Gregg* (Boston: Little, Brown, 1967) and Theodore M. Brown, "Alan Gregg and the Rockefeller Foundation's Support of Franz Alexander's Psychosomatic Research," *Bulletin of the History of Medicine* 61 (1987): 155–82.

3. Robert E. Kohler, "A Policy for the Advancement of Science: The Rockefeller Foundation, 1924–1929," *Minerva* 16 (1978): 480–513. See also Raymond B. Fosdick, *The Story of the Rockefeller Foundation* (New York: Harper & Brothers, 1952), pp. 135–44.

4. Warren Weaver, "The Science of Man," November 29, 1933; RF 3, 915, 1, 6. The phrase was Weaver's. The reorientation of the social sciences had begun even earlier. In 1922, Beardsley Ruml became director of the Laura Spelman Rockefeller Memorial (which later merged with the RF). Under his direction, the Memorial moved away from its practice of appropriating money to welfare organizations and toward the development of a knowledge of human behavior "which in the hands of competent technicians may be expected to result in substantial social control." Ruml is quoted in Franz Samelson, "Organizing the Kingdom of Behavior," *Journal of the History of the Behavioral Sciences* 21 (1985): 39. The Institute of Human Relations at Yale was one product of this new commitment.

5. Staff Report, March 14, 1933; RF 3, 904, 4, 25.

6. Brown, "Alan Gregg," p. 160.

7. *Ibid.*, pp. 161–63.

8. Roger L. Geiger, *To Advance Knowledge: The Growth of American Research Universities, 1900–1940* (New York: Oxford University Press, 1986), p. 166.

9. For example, see Kohler, "A Policy for the Advancement of Science," p. 496.

10. Gerald Jonas, *The Circuit Riders: Rockefeller Money and the Rise of Modern Science* (New York: W. W. Norton Co., 1989), p. 170.

11. Edwin Embree to Raymond Fosdick, August 26, 1925; RF 3, 915, 4, 22.

12. Kohler, "A Policy for the Advancement of Science," p. 500.

13. *Ibid.,* p. 501.

14. Jonas, p. 168.

15. Frank R. Lillie to Wycliffe Rose, June 17, 1924; enclosed in a letter of Lillie to Max Mason, June 5, 1931; RF 1.1, 216D, 8, 104.

16. Lillie to Mason, June 5, 1931; RF 1.1, 216D, 8, 104.

17. Barbara A. Kimmelman, "An Effort in Reductionist Sociobiology: The Rockefeller Foundation and Physiological Genetics, 1930–1942," unpublished manuscript, 1981.

18. Robert E. Kohler, "The Management of Science: The Experience of Warren Weaver and the Rockefeller Foundation Programme in Molecular Biology," *Minerva* 14 (1976): 285.

19. C. C. Little, 9-page enclosure in a letter of February 2, 1937 to Alan Gregg, p. 5; RF 1.1, 200D, 143, 1774.

20. Warren Weaver, "Natural Sciences—Policy: Past Program and Proposed Future Program," extract from agenda for special meeting of trustees, April 11, 1933; RF 3, 915, 1, 6. See pp. 79–80.

21. Warren Weaver, "Progress Report, 1934"; RF 3, 915, 1, 7.

22. Warren Weaver, "Progress Report: The NS Program in Experimental Biology," May 16, 1936, p. 4; RF 3, 915, 1, 8.

23. The significance of Weaver's role in the development of molecular biology has generated much debate. See Kohler, "The Management of Science," and the challenge by Pnina Abir-Am, "The Disclosure of Physical Power and Biological Knowledge in the 1930s: A Reappraisal of the Rockefeller Foundation's 'Policy' in Molecular Biology," *Social Studies of Science* 12 (1982): 341–82. Unfortunately, as Theodore Brown has noted, recent scholarship on the RF has largely ignored Gregg even though his program in psychobiology was a major focus of the Foundation. See Brown, "Alan Gregg," pp. 155–57.

24. Brown, "Alan Gregg," p. 156.

25. "President's Review," RF Annual Report, 1936, pp. 22–23.

26. Weaver, "The NS Experimental Program," pp. 23–25.

27. The report is quoted in Paul Weindling, "The Rockefeller Foundation and German Biomedical Sciences, 1920–40: From Educational Philanthropy to International Science Policy," in Nicolaas A. Rupke, ed., *Science, Politics and the Public Good: Essays in Honour of Margaret Gowing* (London: Macmillan Press, 1988), p. 131.

28. Raymond Fosdick, "Report," RF 1.1, 717, 16, 150. On RF programs in Germany, see also Kristie Macrakis, "The Rockefeller Foundation and German Physics under National Socialism," *Minerva* 27 (1989): 33–57.

29. See memo of Allen Mawrer, principal of University College, October 26, 1934, on "New Scheme of Research in Serological Genetics," RF 1.1, 401, 16, 220.

30. Weaver also supported serological research by A. H. Sturtevant at the California Institute of Technology and Robert Irwin at the University of Wisconsin.

31. Garland E. Allen, "The Eugenics Record Office at Cold Spring Harbor, 1910–1940: An Essay in Institutional History," *Osiris* 2 (1986): 225–64.

32. Excerpt from Alan Gregg's diary, RF 1.2, 200A, 133, 1189.

33. Benno Mueller-Hill, *Murderous Science*, trans. George Fraser (New York: Oxford University Press, 1988), pp. 11, 28–29.

34. Herbert Eugene Walker, *Genetics: An Introduction to the Study of Heredity* (New York: Macmillan, 1913), p. 232.

35. For example, "From one thousand Roumanians today in Boston, at the present rate of breeding, will come a hundred thousand two hundred years hence to govern the fifty descendants of Harvard's sons." William E. Castle *et al.*, *Heredity and Eugenics* (Chicago: University of Chicago Press, 1912), p. 309.

36. Frederick Osborn, "Heredity and Practical Eugenics Today," *Eugenical News* 33 (1948): 5–6.

37. [Frederick Osborn], "Editorial," *Eugenics Quarterly* 1 (1954): 2.

38. C. Nash Herndon, "Heredity Counseling," *Eugenics Quarterly* 1 (1955): 66.

39. Alan Gregg, diary entry, November 17, 1945; RF 1, 401A, 16, 223.

40. A. H. Sturtevant, "Social Implications of the Genetics of Man," *Science* 120 (1954): 405.

41. Amy Sue Bix, "Pavlovian Science Comes to America: Experimental Research of W. Horsley Gantt at the Johns Hopkins University," unpublished manuscript, 1988, p. 14.

42. *Ibid.*, p. 23.

43. *Ibid.*, p. 49.

44. Quoted in Jean Holstein, *The First Fifty Years at the Jackson Laboratory* (Bar Harbor: The Jackson Laboratory, 1979), p. 11.

45. Alan Gregg to C. C. Little, November 12, 1941; RF 1.1, 200D, 143, 1775.

46. Alan Gregg to E. B. Wilson, December 30, 1943; RF 1.2, 200A, 133, 1189.

47. Alan Gregg to C. C. Little, January 3, 1944; RF 1.1, 200, 133, 1189.

48. *Ibid.*

49. C. C. Little to Alan Gregg, January 6, 1944; RF 1.2, 200A, 133, 1189.

50. Robert M. Yerkes to Alan Gregg, March 31, 1944; RF 1.2, 200A, 133, 1189.

51. John Paul Scott, "Investigative Behavior: Toward a Science of Sociality," in D. A. Dewsbury, ed., *Leaders in the Study of Animal Behavior: Autobiographical Perspectives* (Lewisburg, Penn.: Bucknell University Press, 1985), p. 410. Also personal interview with Scott, Bar Harbor, Me., August 27, 1988.

52. Scott, "Investigative Behavior," pp. 396–97.

53. *Ibid.,* pp. 401–402.

54. *Ibid.,* p. 404.

55. *Ibid.*

56. John Paul Scott and John L. Fuller, *Genetics and the Social Behavior of the Dog* (Chicago: University of Chicago Press, 1965), p. 4.

57. Scott, "Investigative Behavior," pp. 396–97.

58. John L. Fuller and William Thompson, *Behavior Genetics* (New York: Wiley, 1960).

59. A list of the forty participants is found in the "Minutes of the Conference on Genetics and Social Behavior," Roscoe B. Jackson Memorial Laboratory, Bar Harbor, Me., September 10–13, 1946. After the conference, Robert Morrison of the RF wrote to Scott suggesting that he try to develop "a strain of obviously schizophrenic dogs" in order to "provide a strong experimental argument for the hereditary nature of the human disease." November 4, 1946; RF 1.2, 200A, 134, 1190.

60. Scott and Fuller made reciprocal crosses and then backcrossed the F1 males to their purebred mothers. They also crossed the same F1 males to their sisters to produce an F2. For further details, see discussions in Scott and Fuller, *Genetics and Social Behavior, passim*, or summary in Scott, "Investigative Behavior," p. 415.

61. Scott and Fuller, *Genetics and Social Behavior,* p. 367.

62. Scott, "Investigative Behavior," p. 416. See also Scott and Fuller, *Genetics and Social Behavior,* p. 388.

63. *Ibid.,* p. 16.

64. Scott, "Investigative Behavior," p. 416.

65. Gregg memo of interview with Little, April 17, 1944; RF 1.2, 200A, 133, 1189.

66. The first motion to appropriate money, on January 19, 1945, also reads: "Much advance must take place in mammalian genetics if we are to approach the study of human heredity wisely. Psychology and psychiatry ignore it and consequently exaggerate the role of experience and environment in the explanation of behavior. Educational policies reflect the same

tendencies to disregard the importance of innate capacities or weaknesses." See RF 1.2, 200A, 133, 1189. The five succeeding motions all include similar comments, as do the RF annual reports.

67. John P. Scott and John L. Fuller, "Heredity and the Social Behavior of Mammals," the Roscoe B. Jackson Memorial Laboratory, 27th Annual Report, 1955–1956, p. 23.

68. "Development of Plans for Twin Study," Rockefeller Archive Center, Population Council, IV, 3B4.2, Box 39, folder 563.

5

A Debate That Refuses to Die

The modern nature-nurture debate began with Sir Francis Galton, a man of many talents who is best known for his pioneering work in statistics and its applications to problems of human heredity. In *Hereditary Genius* (1869), followed five years later by *English Men of Science: Their Nature and Nurture* (1874), Galton argued that in humans just as in brutes, parents and offspring resemble each other. Moreover, this resemblance applies as much to mental as to physical qualities. Natural differences in intelligence, talent, and character explain why some people succeed in life while others fail.

To establish his point, Galton consulted reference works containing biographies of eminent men. He found that high achievement runs in families. Men distinguished in the sciences, arts, and public life were more likely than the public at large to have fathers who were themselves eminent in these fields. Galton considered the possibility that social inheritance explained these results. He acknowledged that "when a parent has achieved great eminence, his son will be placed in a more favourable position for advancement, than if he had been the son of an ordinary person."[1] In the end, however, he discounted the significance of this sort of advantage. It might explain prominence in politics or the military, Galton conceded, but not in science, literature, or the law. Really capable people in these fields would overcome obstacles to success.

In his own day, Galton's views were quite controversial. In the years since, the argument between "hereditarians" and "environmentalists" has gained in sophistication—without coming any closer to a resolution. Perhaps the most striking feature of the nature-nurture debate is the number of times it has ostensibly ended. Indeed, the debate was hardly underway before a victor was announced. In his 1914 presidential address to the British Association, Australia, the geneticist William Bateson asserted that "the long-standing controversy as to the relative importance of nature and nurture . . . is drawing to an end."[2] His was but the first in a long string of similar pronouncements. According to former *Science* editor Daniel Koshland, Jr., it is no longer possible to doubt the "mounting evidence" for the view that both genes and environment contribute to differences in behavior or the relevance of that evidence for public policy. Sensible people reject both the hereditarian claim that genes explain everything and the environmentalist claim that they explain nothing—they stand for a reasonable middle ground between these absurdly simplistic extremes. The controversy is "basically over," Koshland argues, because evidence for an interactionist view is now so strong that even non-scientists (who experience difficulty with "complicated relationships") will have to accept it.[3]

However, the interactionist view is hardly new. In 1911, the geneticist and leading eugenicist Charles Davenport insisted that, "so long as we regard heredity and environment as opposed, so long will we experience endless contradictions in interpreting any trait, behavior, or disease."[4] Indeed, by the 1920s it had become conventional to deny the opposition of nature and nurture, assert that science and common sense had converged on a reasonable middle position, and declare the issue dead. But the recent three-year postponement of a planned conference on genetics and crime tells another story. The conference was aborted when political protests led the National Institutes of Health to terminate its support (which it was later forced to restore on procedural grounds). Critics charged that even to debate the issues was to lend legitimacy to claims of racial differences in predispositions to crime and to promote "a modern-day version of eugenics."[5] The extraordinary attention focused on Richard Herrnstein and Charles Murray's *The Bell Curve*, which spent sixteen weeks on the *New York Times* best-seller list[6] and produced at least two instant books of documents and reviews, likewise testifies to the fact that the many obituaries for the nature-nurture debate have been premature. It seems that this corpse will not stay buried. Why has the debate on the roles of heredity and environment in human behavior been so prolonged, bitter, and apparently intractable? To answer this question, it will help to consider the relationship between political aims and shifting definitions of "hereditarianism" and "environmentalism."

Shifting Meanings

In his 1949 classic, *The Nature-Nurture Controversy*, Nicholas Pastore described the "hereditarian" and "environmentalist" positions as they then were framed. Hereditarians were said to hold that "heredity is more important than environment; individual and group differences are the result of innate factors (either in totality or predominantly); innate characteristics are not easily modified. Where a choice of interpretation is possible, the explanation in genetic terms is the one advanced and favored. To the hereditarian way of thinking, the problem of differential fecundity looms as a most significant one for society."[7] By contrast, environmentalists were said to hold that

> environment is more important than heredity; existing individual and group differences reflect (much more than is commonly thought) differences in opportunity; innate characteristics are easily modified. Furthermore, the "plasticity" of the child is emphasized. Of possible alternative explanations, he chooses the one emphasizing environment. In addition, the environmentalist minimizes the importance of natural inequities in the attainment of success and rejects the eugenic program (as usually conceived).[8]

These positions are illustrated with twenty-four profiles of scientists, half "hereditarians" and half "environmentalists." The hereditarian list is unremarkable. Galton, Bateson, Karl Pearson, Edward East, Paul Popenoe, and Henry H. Goddard would be plausible candidates for any compendium, compiled then or now. But the environmentalist list includes not only such obvious choices as John B. Watson and Franz Boas, but also the geneticists H. J. Muller, J. B. S. Haldane, Lancelot Hogben, and H. S. Jennings. From at least some contemporary standpoints, these geneticists would be labelled "hereditarians." But in 1949, Pastore's categorizations would not have seemed problematic.

Through the 1960s, no geneticist would have dissented from Jennings's view that "all the things that affect character and conduct are deeply influenced by the hereditary materials."[9] *Drosophila* geneticist Jerry Hirsch would in the 1970s become a harsh critic of work on the genetics of human behavior. But in the preceding decade he repeatedly charged social scientists with dogmatic environmentalism. Complaining that "the 'opinion leaders' of two generations literally excommunicated heredity from the behavioral sciences," he called for recognition of both individual and population genetic differences.[10] Richard Lewontin likewise wrote that social scientists had given "insufficient attention" to the work of geneticists, noting that "although there are exceptions, there has been a strong professional bias toward the assumption that human behavior is infinitely plastic, a bias natural enough in men whose professional commitment is to changing behavior."[11]

Hirsch and Lewontin had both been students of Theodosius Dobzhansky. In the 1950s and 1960s, Dobzhansky constantly stressed that the same genotype may be expressed differently in different environments—and that it was thus wrong to equate high heritability with insensitivity to environmental change. He was also an ardent critic of H. J. Muller's eugenics program. (Muller argued that, in civilized societies, the relaxation of natural selection combined with increased exposure to ionizing radiation was producing a dangerous load of deleterious mutations—and that a eugenics program was necessary to counter the trend.)[12] For these reasons, Dobzhansky was considered a leading spokesman for environmentalism. But he was also a founder of modern behavior genetics who believed that "whenever a character variable in human populations has been at all adequately studied, genetic as well as environmental components in its variability have been brought to light. This applies to characteristics of all sorts . . . from skin color, stature, and weight, to intelligence, special abilities, and even to smoking habits."[13]

By the 1970s, the academic landscape had changed. Views that in the prior decade had been considered "environmentalist" now marked one as the opposite. Dobzhansky's favorite maxim was "differences are not deficits." The view that genetic diversity was desirable and should be preserved had once been associated with progressive politics. At least in some circles, it now came to identify one as conservative. Within a few years, and with no acknowledgment of their apparent reversal, some who had chided social scientists for believing that we are Lockean blank slates now assumed that we were (or wrote as though they did). Dobzhansky, among others, was bewildered by the refusal to admit that genes contribute to individual differences in human abilities and aptitudes.[14] But in the politically charged atmosphere of the 1970s, to concede such differences was to risk aligning oneself (or being seen as aligned) with the social views of Arthur Jensen and Richard Herrnstein.

The Debate Is Transformed

In 1969, the psychologist Arthur Jensen claimed in the *Harvard Educational Review* that genetic differences account for at least half of the black-white gap in I.Q. test scores—which explains why compensatory education schemes have failed.[15] The essay produced a storm of controversy. Jensen was initially criticized for exaggerating the significance of heritability estimates and for inappropriately generalizing from statistics on the heritability of I.Q. differences within races to conclusions about differences between them. But even his severest critics took for granted that the differences *within* populations were to some degree heritable, while contesting the generalization to the causes of differences between them.[16]

Two years later, in the *Atlantic Monthly*, Richard Herrnstein published an analogous argument in respect to social class, which he soon expanded to a book, *I.Q. in the Meritocracy*.[17] In the context of a generally radicalized academic environment, the Jensen and Herrnstein studies prompted some scientists to rethink conventional assumptions concerning *individual* differences in intelligence and personality. Within just a few years, the environmentalist position had been transformed.

The nature of that revision was shaped in important ways by the Cyril Burt scandal. In the *The Science and Politics of I.Q.*, the psychologist Leon Kamin charged that Burt's results were, statistically speaking, too good to be true. After reviewing as well all the other classic studies of the heritability of I.Q., which he also found wanting, Kamin concluded that, "there exist no data which should lead a prudent man to accept the hypothesis that I.Q. test scores are in any degree heritable."[18] A heated debate followed on the standards required to demonstrate the heritability of intelligence. Just how rigorous must these standards be?

When the dust settled, it was possible to distinguish essentially two positions. Critics stressed the methodological difficulties involved in designing experiments on the heritability of human mental and behavioral traits, including intellectual performance. They insisted that all existing studies were vitiated by the failure to break the association of genotype and phenotype (a problem resulting from the fact that relatives generally share similar environments). Moreover, for all practical purposes, genetic and environmental effects were too intricately entangled for any genetic contribution to be separated and measured with precision. The enormous efforts required to overcome this problem could not be justified by either the potential scientific or the social interest of the results.

Critics on the political Left had a strong incentive to identify the main issues as methodological. Behavior genetic studies could then be judged and found wanting, not by standards unique to political radicals, but by those of mainstream science. This strategy allowed critics to appeal to a very broad audience: all those who stood for high standards, whatever their politics.

But the focus on the problem of establishing heritability also unwittingly reinforced the view that heritability estimates *mattered*. Moreover, the claim that no heritability had been demonstrated easily elided into the claim that none existed. Although Kamin himself wrote that "there may well be genetically determined differences among people in their cognitive and intellectual 'capacities,'" and noted that the book denied only the heritability of I.Q. test scores, this point was hardly stressed at the time, either by Kamin himself or by other critics of behavior genetic studies.[19]

The argument about heritability might have focused on its scientific value. To the claim of a high within or between-group estimate for some trait, critics might have replied: "So what?" Heritability estimates per se do not establish genetic influence as that is usually understood. For example, there must be some heritability of the differences in black-white I.Q. test scores even if the entire test score gap were explained by social discrimination—since discrimination is linked to (highly heritable) skin color. Or as the philosopher Ned Block has noted, twenty years ago, the heritability of wearing an earring was high. Since virtually only women wore them, the behavior was linked to a chromosomal difference. The heritability of earring wearing must now be much lower. But neither now nor in the past would a heritability statistic provide any causal information about genes.[20] In *The Bell Curve,* Richard Herrnstein and Charles Murray ask: "How Much is IQ a Matter of Genes?" Their answer—"In fact, IQ is substantially heritable"— illustrates the common confusion.[21] Referring to heritability studies, Lawrence Wright likewise writes that "even matters that would seem to be entirely a reflection of one's personal experience, such as political orientation or depth of religious commitment, have been shown by various twin studies to be largely under genetic influence."[22]

Disputed Claims

Given the nature of heritability, it was inevitable that it would finally be established for some trait in some group. Given the nature of the debate, that finding was bound to seem important. The nature of the debate also ensured that people would argue about whether the within or between-group heritability of particular traits was large or small. Indeed, Lawrence Wright notes that "the debate has evolved into a statistical war over percentages—*how much* of our personality or behavior or intelligence or susceptibility to disease is attributable to our genes, as compared with environmental factors."[23] Thus, Mark Snyderman and Stanley Rothman suggest that it is "much safer to argue that a considerable proportion of variation in IQ is due to differences in genes, than to claim that the proportion falls within any given range,"[24] while *Fortune* columnist and author Daniel Seligman asserts that "genetic factors play a major role in explaining IQ differences among individuals,"[25] and Herrnstein and Murray argue that "IQ is substantially heritable."[26] Howard Taylor, on the other hand, claims "that there is no convincing evidence that it is anywhere near substantial."[27]

Even if it were meaningful, debates framed in terms of the relative importance of various factors are notoriously hard to settle. That is one reason why so many controversies in evolutionary biology seem to continue indefinitely or until the major participants die.[28] In his influential book *Inequality,* the sociologist Christopher Jencks suggested that "the real question is not whether such [genetic] differences exist, but whether they are large or trivial."[29] But that question is unlikely to be answered by a statistic. In the (improbable) event that agreement were reached on an estimate of heritability, we would surely debate whether that number was large or small. Is 0.50 a large number? —Is 0.60? —Is 0.40? —Is 0.25?

Moreover, protagonists on both sides of the debate believe—for good reason—that the way the question is settled matters for social policy. In his 1914 address Bateson argued that "all practical schemes for social reform" must be grounded in genetics, and he went on to note that "at every turn, the student of political science is confronted with problems that demand biological knowledge for their solution," in particular, those regarding education and the criminal law.[30] More recently, a leading behavior geneticist similarly argued that we should care about his field since, "some of society's most pressing problems, such as drug abuse, mental illness, and mental retardation, are behavioral problems."[31] From the start, the study of behavior genetics has been promoted for its social utility. Francis Galton aimed to untangle the effects of nature and nurture in order to breed a better race. It was, after all, Galton who in 1883 coined the word "eugenics."

The title of Jensen's 1969 article, "How Much Can We Boost IQ and Scholastic Achievement?" reflected its author's policy concerns, and it was his pessimistic answer that generated controversy and media attention. News accounts of behavior genetic research almost always stress its purported implications. The message is usually that social policy, to be successful, must take genetic differences into account. In commenting on the twin studies of Thomas Bouchard and his colleagues at the University of Minnesota, which purportedly demonstrated a high heritability for such traits as "a penchant for conservatism," a reporter for *U.S. News and World Report* suggested that "psychiatrists and social scientists have long stressed the supremacy of environment in shaping personality and their theories are the basis of many public programs that seek to reverse the social causes of poverty and crime."[32] The implication of the article is clear: these programs have been shown to rest on a naïve belief in the power of environment. The reality, according to Daniel Seligman, is that "most people are no more capable of making it as, say, corporate executives than they are of making themselves into major-league pitchers. Many are incapable of making it even in modestly demanding white-collar jobs."[33] In his view, "genes explain who gets ahead in America, and why."[34] According to Charles Murray as well, bad luck and social obstacles no longer explain why people are poor. Today "what's holding them back is that they're not bright enough to be a physician."[35] *The Bell Curve* has attracted such attention, not because its data or scientific arguments are novel, but because it is an intervention in angry debates about welfare, affirmative action, compensatory education, and immigration policy. Lawrence Wright notes that behavior genetic studies have begun to affect politics and social policy "through the altered understanding of human development which they have engendered." Wright links "the cuts in welfare and job-training programs, the attacks on affirmative action, and the erosion of tax support for public education" to the view that you cannot change people by changing their circumstances.[36] This political dimension of the nature-nurture debate guarantees that it would not end even if the methodological issues were resolved.

Who Speaks for Science?

Protagonists on both sides typically frame the debate in terms of good and bad science. Their opponents are accused of pursuing a political agenda, which explains why they "abuse science." Everyone seems to say (and perhaps truly thinks) that they have been forced to their own conclusions by the weight of the evidence—often adding that the conclusions contradict their prior expectations or personal wishes. Their adversaries, however, have been captured by ideology. "What has happened in the IQ controversy," charge Snyderman and Rothman, "is that the expert voice has been misinterpreted and misrepresented, as science has been perverted for political ends."[37] Commenting on the Bouchard twin studies, Leon Kamin asserted: "This has nothing to do with science. It's a political debate."[38] Of course he means that politics motivates those on the *other* side.

But there is politics everywhere in the nature-nurture debate. There have been political reasons for doing and for attacking the research, for thinking it interesting and for thinking it pointless. That is not necessarily shameful. Both scientists and the public have come to equate political commitment with biased science, and neutrality with objectivity; they think you have to be above the fray to be honest. But it does not follow from having political values that you must fudge your data. Indeed, the claim of neutrality often disguises real interests and motivations.[39]

There is also a deeper sense in which the nature-nurture debate is political. Heritability estimates (which remain the stock in trade of human behavior genetics) necessarily depend on value-laden assumptions about the current social order. As Dobzhansky stressed, since the same genotype may be expressed differently in different environments, heritability cannot be equated with insensitivity to environmental change. That is why plant and animal breeders take as many genotypes as possible and allow them "to develop in as many different environmental dimensions as seems relevant for the species."[40] When it comes to corn, social policy does not depend on which environments are deemed relevant. But when it comes to humans, it does. There are many ways of carving up the existing environment. More important, there are an infinite number of environments that have not yet been tried.

Dobzhansky made this point when he wrote that "invention of a new drug, a new diet, a new type of housing, a new educational system, a new political regime introduces new environments."[41] Given that we can have at best only incomplete knowledge of the "norm of reaction" of any genotype, we cannot predict, for example, how much the intellectual performance of an individual or group might rise in another environment. For this reason, we cannot escape judgments about which manipulations are reasonable. Herrnstein and Murray claim that we do not know how to raise schoolchildren's I.Q. "consistently and affordably."[42] But what constitutes affordability cannot be a strictly scientific judgment.

In his *Science* editorial, Daniel Koshland concluded that "better schools, a better environment, better counseling, and better rehabilitation will help some individuals but not all."[43] The assertion may be true, but it does not follow from the assumption that the relevant behaviors are heritable; it also requires the assumption that we have done all we can or will do to alter the relevant environments. That premise was implicit in Arthur Jensen's claim that black-white differences in I.Q. scores were largely genetic in origin. To make it explicit is to see that the significance of heritability estimates also depends on suppositions about the fixity of existing social arrangements. Many critics of *The Bell Curve* have read its authors as asserting that environmental interventions are futile. But what Herrnstein and Murray actually claim is that the effective interventions are too expensive. On the question of what is worth doing or trying, individuals with different politics are bound to differ. That is why the nature-nurture debate is not simply a matter of good versus bad science.

To represent the controversy as merely a dispute over methods and evidence allows the participants to associate their own position with the cause of science and their opponents' with ideology. Snyderman and Rothman write: "The recent history of this controversy is marked by the increasing subsumption of what is primarily a technical issue . . . under political concerns."[44] But the nature-nurture controversy has never been, and is not now, *simply* a matter of good versus bad science. The views of all the participants are necessarily informed by their politics. To characterize a view as political is not to condemn it. Some scientific controversies have an irreducible political element. That the nature-nurture dispute is one of them explains why it will not soon disappear.

Notes

1. Francis Galton, "Hereditary Talent and Character," *Macmillan's Magazine* 12 (1865): 161.

2. William Bateson, "Presidential Address to the British Association, Australia" (Sydney, Australia, [1914]), reprinted in *William Bateson, F.R.S.: His Essays and Addresses* (New York: Garland, 1984), p. 313.

3. Daniel Koshland, "Nature, Nurture, and Behavior," *Science* 235 (March 20, 1987): 1445.

4. Charles B. Davenport, *Heredity in Relation to Eugenics* (New York: Henry Holt and Co., 1911), p. 252.

5. Juan Williams, "Violence, Genes, and Prejudice," *Discover*, November 1994, p. 95. See also Larissa MacFarquhar, "Take Back the Nitrous," *Lingua Franca*, May–June 1994, pp. 6–7.

6. Joseph D. McInerney, "Why Biological Literacy Matters: A Review of Commentaries Related to *The Bell Curve: Intelligence and Class Structure in American Life*," *Quarterly Review of Books* 71 (1996): 82.

7. Nicholas Pastore, *The Nature-Nurture Controversy* (New York: Kings Crown Press, 1949), p. 14.

8. *Ibid.*

9. H. S. Jennings, *Genetics* (New York: W. W. Norton and Co., 1935), p. 204.

10. Jerry Hirsch, "Behavior Genetics and Individuality Understood," *Science* 142 (1963): 1439. See also *idem*, ed., *Behavior-Genetic Analysis* (New York: McGraw-Hill, 1967); *idem*, "Behavior-Genetic Analysis and Its Biosocial Consequences," *Seminars in Psychiatry* 2 (1970): 89–105.

11. Richard Lewontin, "Race and Intelligence," *Bulletin of the Atomic Scientists* 26 (1970): 5.

12. H. J. Muller, "Our Load of Mutations," *American Journal of Human Genetics* 2 (1950): 111–76. See also Diane B. Paul, "'Our Load of Mutations' Revisited," *Journal of the History of Biology* 20 (1987): 321–35.

13. Theodosius Dobzhansky, "Is Genetic Diversity Compatible with Human Equality?" *Social Biology* 20 (1973): 283.

14. See Diane B. Paul, "Dobzhansky in the 'Nature-Nurture' Debate," in M. Adams, ed., *The Evolution of Theodosius Dobzhansky* (Princeton: Princeton University Press, 1994), pp. 219–31.

15. Arthur Jensen, "How Much Can We Boost IQ and Scholastic Achievement?" *Harvard Educational Review* 39 (1969): 1–123.

16. J. S. Kagan, "Inadequate Evidence and Illogical Conclusions," *Harvard Educational Review* 39 (1969): 274–77; Hirsch, "Behavior-Genetic Analysis and Its Biosocial Consequence," pp. 89–105; Lewontin, "Race and Intelligence," pp. 382–405; and W. E. Bodmer and L. L. Cavalli-Sforza, "Intelligence and Race," *Scientific American* 223 (1970): 19–29.

17. Richard Herrnstein, "I.Q.," *Atlantic* 228 (1971): 63–64; *idem*, *I.Q. in the Meritocracy* (Boston: Little, Brown, 1973).

18. Leon Kamin, *The Science and Politics of I.Q.* (Potomac, Md.: Lawrence Erlbaum, 1974), p. 1.

19. *Ibid.*, p. 176.

20. Ned Block, "Race, Genes, and IQ," *Boston Review of Books,* December 1995, pp. 30–35.

21. Richard J. Herrnstein and Charles Murray, *The Bell Curve: Intelligence and Class Structure in American Life* (New York: Free Press, 1994), p. 105.

22. Lawrence Wright, "Double Mystery," *The New Yorker,* August 7, 1995, p. 48.

23. *Ibid.*

24. Mark Snyderman and Stanley Rothman, *The IQ Controversy* (New Brunswick, N.J.: Transaction Books, 1988), p. 94.

25. Daniel Seligman, *A Question of Intelligence: The IQ Debate in America* (New York: Birch Lane Press, 1992), p. x.

26. Herrnstein and Murray, p. 105.

27. Howard F. Taylor, *The IQ Game: A Methodological Inquiry into the Heredity-Environment Controversy* (New Brunswick, N.J.: Rutgers University Press, 1980), p. 28.

28. See John Beatty, "Weighing the Risks: Stalemate in the 'Classical/Balance' Controversy," *Journal of the History of Biology* 20 (1987): 289–319.

29. Christoper Jencks, *Inequality: A Reassessment of the Effect of Family and Schooling in America* (New York: Basic Books, 1972), p. 72.

30. Bateson, p. 315.

31. Robert Plomin, "The Role of Inheritance in Behavior," *Science* 248 (1990): 184.

32. Julia Reed, "Genes: Little Things That Mean a Lot," *U.S. News and World Report,* December 15, 1986, p. 8.

33. Seligman, p. 35.

34. *Ibid.,* p. 52.

35. Jason DeParle, "Daring Research or 'Social Science Pornography'?" *New York Times,* October 9, 1994, p. 48.

36. Wright, p. 62.

37. Snyderman and Rothman, p. 34.

38. Quoted in Reed.

39. Robert Proctor, *Value-Free Science? Purity and Power in Modern Knowledge* (Cambridge: Harvard University Press, 1991), *passim.*

40. Richard Lewontin, "Genetic Aspects of Intelligence," *Annual Review of Genetics* 9 (1975): 387.

41. Theodosius Dobzhansky, *Evolution, Genetics, and Man* (New York: John Wiley and Sons, 1955), p. 75.

42. Herrnstein and Murray, p. 402.

43. Koshland, p. 1445.

44. Snyderman and Rothman, p. 33.

6

EUGENIC ANXIETIES, SOCIAL REALITIES, AND POLITICAL CHOICES

New Preface

Eugenics has been variously described as an ideal, as a doctrine, as a science (applied human genetics), as a set of practices (ranging from birth control to euthanasia), and as a social movement. The word has been applied to intentions and to wholly unintended effects. It has been defined expansively, to incorporate medical genetics, and narrowly, to wholly exclude it. One goal of this essay was to sort out the confusing complex of meanings and uses.

As originally published, the essay sought to identify two criteria by which eugenics has frequently been marked off from medical genetics: coercive means and social ends. I might also have noted another important but implicit line of demarcation: long-term versus rapid effects. In the early years of genetic counseling, some practitioners argued that the field had nothing to do with eugenics; indeed, its effects, they said, would most likely be dysgenic. Then as now, they equated eugenics with efforts to reduce the incidence of defective genes. But counseling practices, even at their most directive, rarely have this effect. Advising parents to avoid the birth of children with physical or mental disabilities or carriers of the same defective gene not to marry each other reduces the immediate burden of genetic disease—but does nothing to halt the spread of the offending genes. (I discuss this issue more fully in "Eugenic Origins of Medical Genetics" and in "Genes and Contagious Disease: the Rise and Fall of a Metaphor," two essays that appear later in this volume).

In the essay, I also suggested that proponents of genetic medicine tend to favor narrow definitions of eugenics (which disassociate the field from practices that now have ugly connotations), while skeptics favor broad ones (for the opposite reason). I still think that a correct characterization of recent debates. But prior to the 1970s, many medical geneticists employed quite expansive definitions, which are also reappearing. Then and now, they reflect the attitude that there is nothing wrong with eugenics per se although it must be sharply distinguished from terrible versions of the past. In "Eugenic Origins of Medical Genetics" (see pp. 133-156), I argue that, for geneticists at least, the term "eugenics" remained respectable much later than is commonly supposed. Through the 1960s, many geneticists aimed to rehabilitate the term by distinguishing a good (medical) eugenics from the bad eugenics of the past. Philip Kitcher's *The Lives to Come* (1996) exemplifies recent efforts to make the same distinction. In Kitcher's view, once we have genetic knowledge we cannot help practicing eugenics; the issue is whether we do it well or badly.[1] Like the geneticists of the 1950s and 1960s, he employs a broad definition not to condemn genetic testing but to render "eugenics" innocuous. If everything falls under that rubric, the term loses its sting.

I am less sanguine than Kitcher about developments in genetic medicine and also doubt that current efforts to rehabilitate the term will succeed where earlier efforts failed. But I do agree that there is little value in disputing whether a policy really is or is not "eugenics." In "Is Human Genetics Disguised Eugenics?" (not reprinted in this volume but partially incorporated in the revised version of the essay below),[2] I argue that efforts to demarcate eugenics from non-eugenics will prove as fruitless as analogous efforts to demarcate "science" from "non-science" and for the same reason: eugenics, like science, is simply much too heterogeneous.[3]

I believe that disputes about the meaning of eugenics are also unproductive. At present, the term is wielded like a club. To label a policy "eugenics" is to say, in effect, that it is not just bad but beyond the pale. It is a way of ending, not promoting, discussion. It would, in my view, be more useful to ask what scenarios people actually fear when they express anxiety about a eugenics revival, to evaluate which of these scenarios are likely, which possible, and which improbable, and to ask what we can and ought to do to avoid the prospects that are real. To assert that a policy with undesirable effects is also "eugenics" does not add anything substantive to the accusation. What it does add is emotional charge.

In the essay, I express skepticism toward the often-heard claim that the history of eugenics has much to teach us about current dangers in genetic medicine. On reflection, I think I was too dismissive of the value of historical understanding for contemporary policy-making. Indeed, in my recent book *Controlling Human Heredity: 1865 to the Present,* I argue that the history of eugenics elucidates a number of troubling trends.[4] But neither the history nor its lessons are simple. A focus on compulsory sterilization and euthanasia teaches us to be wary of the prospect of genetics in the hands of the state; a focus on eugenicists' advocacy of birth control teaches us that acts are not necessarily benign because their agents are private. Oversimplified accounts of the history of eugenics necessarily obscure the one danger or the other.

Will developments in biomedicine prompt a "new eugenics"? Many people apparently fear that the answer is yes. A host of television programs, trade books, and scholarly and popular articles express their authors' alarm at the prospect of a eugenics revival. As Robert Wright has noted, "Biologists and ethicists have by now expended thousands of words warning about slippery eugenic slopes, reflecting on Nazi Germany, and warning that a government quest for a super race could begin anew if we're not vigilant."[5]

Their message is often buttressed with accounts of the American and German eugenics movements. These histories serve as cautionary tales, meant to remind us that genetics once served corrupt social ends and to alert us to the possibility it may do so again. They also tend to be remarkably similar, down to the slides. They constitute a catalog of the inane and appalling: "Fitter Families" contests at state fairs, immigrants turned back at Ellis Island,

Nazi death camps. The moral is clear, if vague in respect to details. Genetics has been badly abused in the past. If we are not careful, it may happen again. Thus, precautions need to be taken.

Perhaps understanding the past can help us chart a course through the dangerous shoals of genetic engineering although, as Michael Lockwood suggests, "that may be a little like looking to the history of the temperance movement for guidance on the contemporary problems of drug addiction."[6] It is in any case highly improbable that these canned histories, with their attitude of total contempt for the past, will promote a critical perspective on current developments in biomedicine. They are in fact far less likely to inspire self-criticism than self-congratulation. In comparison with our grandparents, who held inexplicably absurd and odious ideas, we look pretty good.[7] Indeed, making us feel good may be among their chief functions.

Consider the ritual injunctions to "be vigilant." What exactly are we to guard against? Few commentators believe that a new eugenics will simply repeat the mistakes of the past. We might therefore expect to be told in what guises eugenics actually appears today, or will likely appear tomorrow. However, these discussions are typically abstract and general. Concrete cases would of course provoke controversy. Thus we are cautioned to be on guard—against nothing in particular. The lessons of history turn out to be vacuous.

Perhaps the unthreatening character of the discussion explains why advocates as well as critics of the Human Genome Initiative acknowledge eugenics as a serious concern. Of course not everyone does. Some simply dismiss eugenic anxieties, while others view anxiety itself as the problem.[8] But the literature celebrating the project is replete with homilies about the need for vigilance lest abuses recur. Even James D. Watson, its first and very hard-headed director, thought the eugenics issue was real.[9] It has become, in effect, the "approved" project anxiety.

In comparison, other concerns receive short shrift. The past two years have witnessed publication of a raft of popular books celebrating the genome initiative. Nearly all focus on its promise for medicine.[10] None questions whether mapping and sequencing the whole human genome represents a cost-effective way to prevent or cure disease. Their authors warn of the need to guard against a resurgence of eugenics. They do not say what they mean by eugenics or why it is bad. The answers to both questions are generally believed to be obvious. The burden of this essay is to show they are not—and to explore some political consequences of our failure to explore these assumptions.

The Multiple Meanings of Eugenics

Eugenics is a word with nasty connotations but an indeterminate meaning. Indeed, it often reveals more about its user's attitudes than the policies, practices, intentions, or consequences labelled. (The problem of multiple meanings was recognized by the Commission of the European Communities when it omitted the word "eugenics" from its revised human genome analysis proposal on the grounds that it "lacks precision."[11]) To oppose eugenics is to signal that one is socially concerned and morally sensitive but it does not predict one's stance on any particular issue. In 1990, the International Huntington Association and the World Federation of Neurology adopted guidelines, based on the recommendations of a joint committee, for the use of predictive genetic tests. The committee considered the refusal to test women who "do not give complete assurance that they will terminate a pregnancy where there is an increased risk" of Huntington's disease to be acceptable policy.[12] Was the committee endorsing eugenics? Some would say yes, while most its members would certainly be appalled by the suggestion.

The superficiality of public debate on eugenics is partly a reflection of these diverse, sometimes contradictory meanings, which result in arguments that often fail to engage. Thus, for some "eugenic" may imply a social intention; to others it may also describe consequences that are *unintended*. From the latter's standpoint, it may make sense to call the practice of abortion following prenatal screening "eugenics." From the former's, it does not. Few, if any, women choose abortion with the aim of improving the gene pool. But private decisions may, taken collectively, have population effects. These consequences would appropriately be labelled eugenic (or perhaps dysgenic) given some definitions—and equally inappropriate given others.[13] And that is but one source of confusion.

Francis Galton, who coined the term in 1883, defined eugenics as "the science of improvement of the human race germ plasm through better breeding" and as "the study of agencies under social control which may improve or impair the racial qualities of future generations."[14] Both versions identify eugenics as science, rather than practice. But if eugenics only implied *study,* it would hardly arouse such indignation. Some modern definitions are even more innocuous. Eugenics is "the concern with the genetic improvement of mankind";[15] "the attempt to improve the population through selective breeding";[16] "the promotion of reproductive options favoring desired human genetic traits, especially health, longevity, talent, intelligence, and unselfish behavior;"[17] "attempts to improve hereditary qualities through selective breeding."[18] Definitions this broad will necessarily incorporate medical genetics. They may even incorporate ordinary acts of human reproduction.[19]

Given eugenics' bad reputation, some people prefer much narrower definitions. Those definitions employed by medical geneticists usually identify eugenics with a social aim or coercive means. For example, a participant at the 1991 International Congress of Human Genetics asserted: "Eugenics presumes the existence of significant social control over genetic and reproductive freedoms. Genetics does not require any special control over genetics or reproductive freedom."[20] The criteria of social control and social purpose are often combined, as in the definition of eugenics as "any effort to interfere with individuals' procreative choices in order to attain a societal goal."[21]

Critics of genetic medicine, on the other hand, tend to employ very broad definitions. Thus Abby Lippman charges: "Though the word 'eugenics' is scrupulously avoided in most biomedical reports about prenatal diagnosis, except where it is strongly disclaimed as a motive for intervention, this is disingenuous. Prenatal diagnosis presupposes that certain fetal conditions are intrinsically not bearable."[22] Some commentators have warned of a new "homemade" or "back door" eugenics arising from individual choices. In their view, the real danger arises not from state policy but from our increased capacity to *choose* the kind of children we want.[23] According to Troy Duster: "When eugenics reincarnates this time, it will not come through the front door, as with Hitler's *Lebensborn* project. Instead, it will come by the back door of screens, treatments, and therapies."[24] These practices will seem to enhance health, and most likely be welcomed. Implicit in these critical analyses is an expansive definition of eugenics. Both broad and narrow definitions serve a clear political purpose. The former associate genetic medicine with odious practices and thus arouse our suspicions; the latter dissociate it from these practices and thus reassure.

Eugenics and Coercion

It is sometimes said that what people object to in eugenics is not the goal, such as improving the health of the population, but the means employed to achieve it.[25] From this standpoint, in the absence of coercion (as reflected in law or obvious forms of social pressure), policies designed with the good of the population in mind are not properly labeled "eugenic." But there are also problems, both historical and analytical, with this approach to defining eugenics. Many eugenicists stressed the voluntary character of their proposals. This was especially true in Britain. For example, the English socialist Havelock Ellis insisted that "the only compulsion we can apply in eugenics is the compulsion that comes from within."[26] It likewise excludes most "positive eugenics" (which aims to increase desirable traits rather than reduce undesirable ones), since these schemes are usually voluntary. Thus, H. J. Muller's proposal to artificially inseminate women with the sperm of particularly estimable men would not, in this perspective, qualify as eugenics. Nor would William Shockley's similar project involving Nobel Prize winners.

Moreover, it is no simple matter to determine whether a policy is coercive, and indeed there is no value-neutral way to decide. Coercion has different meanings in different political traditions; to classical liberals and contemporary (libertarian) conservatives, it "implies the deliberate interference of other human beings" with actions a person would otherwise take.[27] However, to liberals in the tradition of T. H. Green or John Dewey, or to socialists, coercion is not simply a matter of removing formal and legal barriers: we are free to choose only when we have the practical ability to agree or refuse to do something. From their standpoint, a *situation* may also be coercive. In the former tradition, the potential parents of a severely disabled child are considered free to abort the fetus or bring it to term. In the latter, they may not be, given the enormous medical and other costs of caring for such a child. On the view of liberals or socialists, parents could be coerced into aborting a fetus by the threatened loss of insurance coverage or lack of social services. (This is not to suggest that pressure would evaporate with national health insurance. Even in a socialized system, potential parents who lack confidence in their society's willingness to care for their child once they are no longer able do so, "may choose to

terminate a pregnancy against their own wishes and beliefs."[28])
The existence of these pressures is wholly compatible with a commitment to reproductive autonomy—on the dominant understanding of that principle. As Ruth Chadwick has noted, autonomy has
come increasingly to be identified with self-reliance; "standing on
your own two feet." That interpretation explains why the same
people can vote to cut social services while upholding the principle
of freedom of reproductive choice.[29]

Our language obscures the fact that there are different under-
standings about what it means to act autonomously and choose
freely. We speak of "private, voluntary" choice as though individual
choices are *ipso facto* free.[30] But this is only true if we define freedom as the absence of legal restraint. In some political traditions,
whether parents are considered free to bring a severely disabled
fetus to term is a matter not just of law but of economics.

Disputes about the uses of new genetic technologies cannot be
easily resolved since they are linked to more fundamental values,
about which people disagree. In the literature on social and ethical
implications of biomedicine, we often find dictates such as "coercion
should not be used."[31] Such injunctions miss the point. There is
already general agreement that coercion is bad; the problem is a
lack of agreement on what coercion *is*.

The Nazi euthanasia program and many of the American state
sterilization statutes were obviously compulsory. But some anxieties about coercion extend beyond the possibility of being segregated or sterilized or shot by the state. They include concerns about
lacking realistic alternatives to the decision to undergo predictive
genetic testing or to abort a genetically imperfect fetus as a consequence of policies adopted by employers, insurers, or health care
providers who want to save money (or in the case of the last,
protect against malpractice suits). After all, "avoiding the conception of an infant at risk for a genetic disease—or avoiding the birth
of a fetus prenatally diagnosed as having one—will often be less
expensive than clinical management."[32]

In the absence of public policy designed to prevent it, reproductive decisions will often be driven by the conjoined interests of
powerful nonstate entities. As Dennis Karjala notes, "given the
natural revulsion that most people feel for interference through
mandatory testing or, even worse, mandatory abortion, the issues
[of 'genetic freedom and genetic responsibility'] are likely to be

raised obliquely" through the health insurance system, HMO policies, or doctor pressure.[33] In some political traditions, the exercise
of this kind of social power calls into question the voluntariness of
a woman's choice. In others, it does not. Thus, the apparent social
consensus on the value of reproductive freedom and "autonomous"
decision making dissolves once we begin to probe a bit.

Eugenics and Social Aims

Sometimes the rationale, rather than the means employed, identifies
a policy as eugenic. Programs are often tagged with the label if
their intent is to further a social or public purpose, such as reducing costs born by the sociomedical system or sparing future generations suffering. In this perspective, genetic counseling or support
for biomedical research motivated by concern for the quality of the
"gene pool" would be eugenic while the same practices motivated
by the desire to increase the choices available to individuals would
not. This same criterion is also employed to distinguish the old
eugenics from the new.

In his important book *In the Name of Eugenics*, Daniel Kevles
characterized postwar advances in medical genetics and biomedical
engineering as a "new eugenics."[34] The phrase was bound to be
provocative. After World War II, eugenics fell increasingly into disrepute. But many geneticists remained convinced that improving
the biological quality of human populations was a worthy goal and
feared it would be abandoned in the backlash against establishment eugenics. They therefore searched for new and politically
neutral ways to pursue their objectives. From these efforts emerged
the field of genetic counseling and ultimately such medical techniques as prenatal diagnosis and experiments in gene therapy. Many
who approve these developments also abhor "eugenics." Hence the
controversial character of Kevles's expression.

It was not, however, Kevles's intent to condemn these new
techniques and therapies. Indeed, he is convinced that the "new
eugenics" has jettisoned not only the social prejudices that marked
much of the old, but the social *interests* that spurred the first generation of medical geneticists. In his view, some who developed
(and funded) the field of medical genetics may have aimed to improve the quality of the gene pool. But in the 1960s, the ethos of

genetic counseling shifted from concern with the welfare of populations to the welfare of individual families, as determined by families themselves. This change marks a welcomed, decisive break with the past. Previous abuses resulted, in his words, from elevating "abstractions—the 'race,' the 'population,' and more recently the 'gene pool'—above the rights and needs of individuals and their families."[35] In effect, old (bad) eugenics reflects interests that are social; new (at least potentially good) eugenics reflects interests that are private and individual.

But there are at least two problems in employing the individual/social criterion to distinguish either good eugenics from bad or what is genuinely eugenic from what is not. First, the criterion requires a knowledge of motives, which may not be obvious and are often mixed. Indeed, genetic services are generally justified on one of two very different grounds: that they increase the options available to women and/or that they reduce the incidence of genetic disease. If a social purpose is to be the litmus test of eugenics, we must assess the relative importance of different aims, which are not always made explicit—and when they are, may disguise the truth. Sensitivities about abortion provide a strong incentive to defend prenatal testing in the language of choice rather than the language of finance.

Second, private acts do have social effects. And it at least requires argument to show why social consequences should not be a matter of social concern. The conventional counterpoising of the interests of individuals (which are good) to those of society (which are bad, or at least less compelling) ignores the fact that society is constituted by other individuals, and that individuals have social interests. For historical reasons, we have come to take for granted that the rights and needs of individuals (or individual families) should take precedence over rights and needs that are social. But these alternatives are more complex than current discussion admits. It is not so clear that all good is on the side of the individual; indeed, it is not so clear what it means to be on the side of the individual in the first place.

From a historical standpoint, the desire to draw the line here is certainly understandable. Whether or not we can agree on what eugenics really is, there were policies associated with the eugenics movement that we now find abhorrent. And these policies were defended on the grounds that individual desires must be sacrificed to a larger public good.

In the extreme view, the individual was thought to count for nothing, the larger community all. The following passage from Madison Grant's influential *The Passing of the Great Race* (1916) illustrates this perspective:

> Mistaken regard for what are believed to be divine laws and a sentimental belief in the sanctity of human life tend to prevent both the elimination of defective infants and the sterilization of such adults as are themselves of no value to the community. The laws of nature require the obliteration of the unfit, and human life is valuable only when it is of use to the community or race.[36]

Few eugenicists spoke in so harsh a language. But the need for individual sacrifice on behalf of the larger social good was a belief shared by all eugenicists, whatever their other social and economic views. Thus, the socialist Lancelot Hogben, whose views were generally shared by those on the Marxist or Fabian left, wrote that, "The belief in the sacred right of every individual to be a parent is a grossly individualistic doctrine surviving from the days when we accepted the right of parents to decide whether their children should be washed or schooled."[37]

Because the *raison d'être* of eugenics was the sacrifice of individual desire to public good, it was often characterized as a "secular region." The idea that science could, and should, function as a religion was first proposed by Francis Galton and later defended by various socialists such as George Bernard Shaw.[38] Bertrand Russell also argues the point in his essay on eugenics in *Marriage and Morals:*

> The idea of allowing science to interfere with our intimate personal impulses is undoubtedly repugnant. But the interference involved would be much less than that which has been tolerated for ages on the part of religion. Science is new in the world, and has not yet that authority due to tradition and early influences that religion has over most of us, but it is perfectly capable of acquiring the same authority and of being submitted to with the same degree of acquiescence that has characterized men's attitude toward religious concepts.... Religion has existed since before the dawn of history, while

science has existed for at most four centuries; but when science has become old and venerable it will control our lives as much as religion has ever done.[39]

Grant, Hogben, and Russell had little in common except their enthusiasm for an alliance of science and state that today seems at least naïve. Few people are any longer inclined to celebrate either. Indeed, the notion that individual desires should sometimes be subordinated to a larger social good has itself gone out of fashion, to be replaced by an ethic of radical individualism. In 1885, Jane Clapperton's assertion that the socialist state would eventually have to restrain the sexuality of those "who persist in parental action detrimental to society" was not especially controversial.[40] Some people still talk this way. Margery Shaw, past president of the American Society of Human Genetics, answers "no" to the question of "whether of not a defective fetus should be allowed to be born," expressing optimism that "parental rights to reproduce will diminish as parental responsibilities to unborn offspring increase."[41] But hers is now an unfashionable minority position. And even Shaw would rely on tort liability—not legislation—to enforce parental "responsibilities."

In the late 1980s, Dorothy Wertz and John Fletcher queried medical geneticists in nineteen countries about their attitudes toward ethical problems in genetic counseling, prenatal diagnosis, and screening. They found autonomy to be a dominant value. More than ninety percent of geneticists in the United States, and more than eighty percent in other countries, believe that counseling should be non-directive.[42] A statistic from another recent study by Wertz, Fletcher, and John Mulvihill provides an even more striking illustration. In 1972–73, only one percent of genetic counselors in the United States would perform prenatal diagnosis (or would refer parents) for selection of fetal sex in the absence of X-linked disease. In 1975, the figure was twenty-one percent. In 1985, it was sixty-two percent.[43] And in a recent survey, Deborah Pencarinha and colleagues found that eighty-five percent of masters-level genetic counselors would either counsel or refer patients desiring sex-selection.[44]

There are complex reasons for this ideological shift. One is the reaction to abuses committed in the name of eugenics. Consider the history of attitudes toward sterilization of retarded persons. In the 1927 case of *Buck v. Bell,* which upheld the Virginia sterilization statute, Justice Oliver Wendell Holmes wrote: "It is better for all the world, if instead of waiting to execute degenerate offspring for crime, or to let them starve for their imbecility, society can prevent those who are manifestly unfit from continuing their kind. The principle that sustains compulsory vaccination is broad enough to cover cutting the Fallopian tubes."[45]

The 1981 *Grady* decision provides a striking contrast with both *Buck v. Bell* and the case of Karen Ann Quinlan. In *Grady,* the parents of an adolescent girl with Down syndrome feared that their daughter, who was about to enter a sheltered workshop, would be raped or seduced. They asked their physician to sterilize her, but the local hospital refused to permit it. The parents then asked the court to authorize the procedure. When the judge ruled in their favor, the New Jersey attorney general appealed to the state Supreme Court, which confirmed the lower court decision. But in contrast to the *Quinlan* case, where it ruled that the decision to terminate life support systems could be delegated to family and physicians, the New Jersey Supreme Court held that sterilization of retarded persons always required judicial approval. The court explained its departure from *Quinlan* by reference to the "sordid history" of eugenic sterilization.[46]

A related factor is certainly the expanding jurisprudence of privacy, which has centered on sexuality and procreation. As early as 1942, the court unanimously overturned a sterilization law in an opinion that termed procreation "one of the basic civil rights of man."[47] The court further expanded the scope of privacy and reproductive freedom in *Griswold v. Connecticut* (1965), where it struck down a law prohibiting the use of contraceptives, in *Eisenstadt v. Baird* (1972), where it held that "if the right of privacy means anything, it is the right of the individual, married or single, to be free of unwarranted governmental intrusion into matters so fundamentally affecting a person as the decision whether to bear or beget a child,"[48] and of course in *Roe v. Wade* (1973). The repudiation of the ethos underlying *Buck v. Bell* could hardly be more complete. This reversal has been the joint product of a liberal Supreme Court and broader social movements, in particular feminism.[49]

But the claim that women have an absolute right to control their bodies also has implications that make some feminists uneasy. As Elizabeth Fox-Genovese has recently suggested, contemporary feminist theory is marked by "the uneasy coexistence of

communitarian and individualistic commitments."[50] Tensions arising from these dual commitments are especially evident in discussions of prenatal diagnosis for sex selection. Many feminists
(including genetic counselors, most of whom are women) are made
very uncomfortable by this practice. After all, as Dorothy Wertz
and John Fletcher note, "gender is not a disease."[51] But if women
have an absolute right to reproductive choice, they cannot consistently be denied the right to choose the sex of their offspring. And
if it is acceptable to choose fetal sex, on what grounds can choice
be denied to parents who want to give their child a competitive
advantage with respect to intelligence, height, or other socially
desired characteristics? It is a question that opens troubling vistas
to many feminists.

Another strand in the tapestry of arguments for reproductive
autonomy can be traced to Carl Rogers, father of "client-centered
counseling," whose theories strongly influenced the ethic of genetic
counseling in the 1970s.[52] From the 1940s (when heredity clinics
were first established) through the 1960s, genetic services were generally provided as a sideline by physicians and research-oriented
Ph.D. geneticists. It was only in 1969 that Sarah Lawrence College
established the first Master's level program for professional counselors. Thus, in North America the vast expansion in genetic services of the 1970s was accompanied by a shift both in the gender
and training of counselors. The new service providers were much
more attuned to the psychological dimensions of the counseling
process. They were influenced in particular by Rogers's view of
the counselor's role as clarifying and objectifying the client's own
feelings.[53]

Reproduction, the State, and the Market

The convergence of these forces in the 1970s ensured that, in respect to reproduction, autonomy would supersede other competing
values. But the commitment to personal autonomy has obscured
the fact that individual reproductive decisions do have social consequences and that in a market system the privatization of those
decisions will result in their commoditization. Robert Nozick's concept of a "genetic supermarket," a system that allows prospective
parents to order (within limits) the genetic characteristics of their

offspring, is the logical outcome of this trend, in which the power of the market ultimately replaces that of the state.[54] As Robert Wright suggests, the real problem is not the one we most fear: a government program to breed better babies. "The more likely danger," he writes, "is roughly the opposite; it isn't that the government will get involved in reproductive choices, but that it won't. It is when left to the free market that the fruits of genome research are most assuredly rotten."[55] He also notes (following Arthur Caplan) that those who now take advantage of prenatal screening are concentrated at the upper end of the income scale while children with Down syndrome are almost certainly being born disproportionately to those at the bottom.[56]

Wright calls this development "homemade eugenics." I do not think that labelling it eugenics helps to clarify the real issues it raises. The social structures and processes involved are not those envisaged in the eugenics literature or implicated in its historical practice. The word functions here, as it often does, to mobilize anxieties. It says: no right-thinking person could fail to object to the practice described. It is used, in effect, as a club. But the problem identified by Wright (and Duster) is certainly real. And it is one we will have a hard time thinking through, much less resolving politically, for those who are most concerned with these particular (mis)uses of genetics tend also to the most committed to be the principle of reproductive autonomy.

We have essentially retreated to a position associated with nineteenth-century liberalism: that there are two spheres of activity— the one in which the individual possesses absolute liberty; the other in which society might legitimately interfere. That distinction was originally asserted by John Stuart Mill. He thought there were two kinds of actions, those with and without social consequences. Where one's actions affect others, society might intervene. But "over himself, over his own mind and body, the individual is sovereign."[57]

Few philosophers think this distinction workable; it is difficult to identify any activities devoid of social effects. But it is perhaps worth noting that reproduction and parenting were almost the only activities in respect to which Mill urged *greater* state responsibility. He wrote:

The fact itself, of causing the existence of a human being, is one of the most responsible actions in the range of human life.

> To undertake this responsibility—to bestow a life which may be either a curse or a blessing—unless the being on whom it is to be bestowed will have at least the ordinary chances of a desirable existence, is a crime against that being.[58]

Thus, even Mill, who wished to grant the widest possible scope to individual action, and the least to society, thought reproduction and parenting inherently social.[59]

As the passage from Mill suggests, reproductive decisions have—at a minimum—consequences for another person. Our language conceals this fact. For example, we talk of "individual families" as though the interests of parents and children are necessarily identical. Yet potential conflicting interests within families are reflected in "Baby Doe" cases (where parents want their infants to die) and in the cases of parents who petition the courts to approve sterilization of retarded (usually female) children.

We do recognize potential conflicts in non-reproductive spheres. Thus, in respect to the issues of child labor or child abuse we grant that the rights of individuals (as exercised by some individuals) may threaten the interests or freedom of other individuals in the absence of some intervening social choice. Consider the latter case. While ignoring the interests of the child, what if it could be shown with sufficiently persuasive evidence that violent and abusive treatment of children produced violent and abusive adults? Would there not then be a compelling argument for prohibiting child abuse *in addition* to an argument from the defense of the child's rights? It is commonly accepted that there is a social interest of this kind in education, and that means the interest not of some abstraction called "Society," but of other individuals who constitute society.

Unlike the child abuse situation, however, individuals will certainly differ in their evaluation of the social effects of reproductive decisions. Whether they are viewed as good or bad, important or unimportant, is a function of deeply held values, about which people disagree. Any social policy will necessarily favor some values over others, and thus engender potentially bitter social conflict. Such conflict has largely been avoided through the privatization of reproductive decisions. But when the scope of politics is reduced, that of the market is usually expanded, thus replacing one form of social power with another.

The issues raised by contemporary developments in biomedicine are enormously difficult. There is, alas, no algorithm for their solution. An ethic of radical individualism might insulate reproductive decisions from racist and reactionary forces that dominated the eugenics movement and permit potentially bitter social conflicts to be suppressed. But it also leaves society, as John Dunn has remarked of nineteenth-century liberals, "confronting History in the guise of an ostrich," without any way even of thinking through the issues involved.[60] And this refusal to consider the social implications of the commitment to absolute reproductive autonomy has a consequence even radicals have been reluctant to face: to retreat from politics is ultimately to embrace the market.

Notes

1. Philip Kitcher, *The Lives to Come: The Genetic Revolution and Human Possibilites* (New York: Simon and Schuster, 1996).

2. Diane B. Paul, "Is Human Genetics Disguised Eugenics?" in Robert F. Weir *et al.*, eds., *Genese and Human Self-Knowledge: Historical and Philosophical Reflections on Modern Genetics* (Iowa City: University of Iowa Press, 1994), pp. 67–83.

3. On the history of demarcation criteria in respect to science, see Larry Laudan, "The Demise of the Demarcation Problem," in Michael Ruse, ed., *But Is It Science?* (New York: Prometheus Books, 1988), pp. 337–50.

4. Diane B. Paul, *Controlling Human Heredity: 1865 to the Present* (Atlantic Highlands, N.J.: Humanities Press, 1995).

5. Robert Wright, "Achilles' Helix," *New Republic,* July 9 & 16, 1990, p. 27.

6. Michael Lockwood, "The Improvement Movement," *Nature*, October 10, 1985, p. 481.

7. Charles Rosenberg notes that "eugenics has become a familiar term to historians and informed readers, but debate has on the whole stimulated more posturing and self-congratulatory moralizing than serious scholarship." Review of Daniel J. Kevles, *In the Name of Eugenics,* in *Journal of American History* 73 (June 1986): 232. A number of more nuanced accounts have appeared since Rosenberg's review. See particularly the collection of essays edited by Mark Adams, *The Wellborn Science: Eugenics in Germany, France, Brazil, and Russia* (New York: Oxford University Press, 1990), and Richard A. Soloway, *Demography and Degeneration: Eugenics*

and the Declining Birthrate in Twentieth-Century Britain (Chapel Hill: University of North Carolina Press, 1990).

8. Daniel E. Koshland, Jr. dismisses the issue in his editorial "Sequences and Consequences of the Human Genome," *Science*, October 13, 1989, p. 189, and in his reply to Salvador Luria, *Science*, November 17, 1989, p. 270. Nancy Wexler suggests that "it's critical that people appreciate the limits of what can be done, so that their fears don't intrude on the benefits that could come out of genetic research." Quoted in William Saletan, "Genes 'R Us," *New Republic,* July 16 and 24, 1989, p. 18.

9. At the 1989 Human Genome I Conference, Watson said: "We have to be aware of the really terrible past of eugenics, where incomplete knowledge was used in a very cavalier and rather awful way, both here in the United States and in Germany." Quoted in Joel Davis, *Mapping the Code* (New York: Wiley, 1991), p. 262. See also James D. Watson, "The Human Genome Project: Past, Present, and Future," *Science*, April 6, 1990, pp. 244–48.

10. For example, Joel Davis, *Mapping the Code* (New York: Wiley, 1991); Lois Wingerson, *Mapping our Genes* (New York: Dutton, 1990); Jerry E. Bishop and Michael Waldholz, *Genome* (New York: Simon and Schuster, 1990).

11. "Modified Proposal for a Council Decision Adopting a Specific Research and Technological Development in the Field of Health: Human Genome Analysis (1990–1991)," *Official Journal of the European Communities,* CB-CO-89-485-EN-C, Brussels (November 13, 1989), p. 3.

12. "Ethical Issues Policy Statement on Huntington's Disease Molecular Genetics Predictive Test," *Journal of Medical Genetics* 27 (1990): p. 37.

13. This double meaning of eugenics is discussed by Elof Axel Carlson in "Ramifications of Genetics," *Science*, April 25, 1986, p. 531.

14. Galton first coined the term eugenics (from the Greek for "well-born") in his 1883 book, *Inquiries into the Human Faculty and Its Development* (New York: Macmillan).

15. Mark H. Haller, *Eugenics: Hereditarian Attitudes in American Thought* (New Brunswick, N.J.: Rutgers University Press, 1984), p. ix.

16. Wingerson, *Mapping Our Genes,* p. 304.

17. Elof Axel Carlson, *Human Genetics* (Lexington, Mass.: D. C. Heath and Company, 1984), Glossary v.

18. Davis, *Mapping the Code,* p. 283.

19. Carlson, "Ramifications of Genetics," p. 531.

20. F. D. Ledley, "Differentiating Genetics and Eugenics on the Basis of Fairness," poster 1818, Eighth International Congress of Human Genetics, Washington, D.C., October 6–11, 1991.

21. Neil Holtzman, *Proceed with Caution* (Baltimore: Johns Hopkins University Press, 1989), p. 223.

22. Abby Lippman, "Prenatal Genetic Testing and Screening: Constructing Needs and Reinforcing Inequalities," *American Journal of Law and Medicine* 17 (1991): 24–25.

23. Wright, p. 27.

24. Troy Duster, *Backdoor to Eugenics* (New York: Routledge, 1990), p. x.

25. Thus, the authors of an Office of Technology Assessment analysis of the Genome Project write: "It may well be that the problem with positive eugenics has more to do with the means than with the ends. The basic objective of improving the human condition is generally supported, although debates about just what constitutes such improvement continue. Many concerns about eugenic policies in the past focused on the methods used to obtain them, such as sterilization, rather than on the ends themselves." See *Mapping Our Genes—Genome Projects: How Big, How Fast?* (Washington, D.C.: U.S. Congress, OTA-BA-373, April, 1988), p. 85.

26. Havelock Ellis, *The Task of Social Hygiene* (London: Constable, 1912), pp. 45–46.

27. Isaiah Berlin, "Two Concepts of Liberty," reprinted in *Four Essays on Liberty* (New York: Oxford University Press, 1969), p. 123.

28. Angus Clarke, "Genetics, Ethics, and Audit," *Lancet* 335 (May 12, 1990): 1146.

29. Ruth Chadwick, "What Counts as Success in Genetic Counselling?" *Journal of Medical Ethics* 19 (1993): 43–46.

30. For a related discussion of the ways in which the discourse of choice functions to disguise social reality, see Abby Lippman, "Prenatal Genetic Testing and Screening: Constructing Needs and Reinforcing Inequities," *American Journal of Law and Medicine* 17 (1991): 15–50.

31. For example, see David Suzuki and Peter Knudtson, *Genethics: The Clash Between the New Genetics and Human Values* (Cambridge, Mass.: Harvard University Press, 1989), pp. 160, 346.

32. Neil A. Holtzman and Mark A. Rothstein, "Eugenics and Genetic Discrimination," *American Journal of Human Genetics* 50 (March 1992): 457.

33. Dennis J. Karjala, "A Legal Research Agenda for the Human Genome Initiative," *Jurimetrics* 32 (Winter 1992): 159.

34. Daniel J. Kevles, *In the Name of Eugenics: Genetics and the Uses of Human Heredity* (New York: Knopf, 1985).

35. *Ibid.*, p. 301.

36. Madison Grant, *The Passing of the Great Race* (New York: Charles Scribner's, 1916), pp. 44–45.

37. Lancelot Hogben, *Genetic Principles in Medicine and Social Science* (London: Williams and Norgate, 1931), p. 207.

38. Shaw wrote that "there is no now reasonable excuse for refusing to face the fact that nothing but a eugenic religion can save our civilization." See his *Sociological Papers* (London: Macmillan, 1905), pp. 74–75.

39. Bertrand Russell, "Eugenics," in *Marriage and Morals* (London: G. Allen and Unwin, 1924), pp. 272–273.

40. Jane Clapperton, *Scientific Meliorism* (London: K. Paul, Trench, 1885), p. 102. Quoted in Soloway, *Demography and Degeneration,* p. 102.

41. Margery W. Shaw, "To Be or Not to Be? That Is the Question," *American Journal of Human Genetics* 36 (1984): pp. 1, 9. See also her essays, "The Potential Plaintiff: Preconception and Prenatal Torts," in A. Milunsky and G. Annas, eds., *Genetics and the Law II* (New York: Plenum Press, 1980), and "Conditional Prospective Rights of the Fetus," *Journal of Legal Medicine* 5 (1984): 63–116.

42. Dorothy C. Wertz and John C. Fletcher, eds., *Ethics and Human Genetics: A Cross-Cultural Perspective* (New York: Springer-Verlag, 1989), pp. 26–31.

43. Dorothy C. Wertz, John C. Fletcher, and John J. Mulvihill, "Medical Geneticists Confront Ethical Dilemmas: Cross-cultural Comparisons among 18 Nations," *American Journal of Human Genetics* 46 (1990): 1209–10.

44. Deborah Pencarinha *et al.,* "A Study of the Attitudes and Reasoning of M.S. Genetic Counselors Regarding Ethical Issues in Medical Genetics," Poster Presentation, 12th International Congress of Human Genetics, Washington, D.C., October 6–11, 1991.

45. J. Cynkar, *"Buck v. Bell:* 'Felt Necessities' v. 'Fundamental Values'?" *Columbia Law Review* 81 (1981): 1418–61.

46. Philip R. Reilly, *The Surgical Solution: A History of Involuntary Sterilization in the United States* (Baltimore: Johns Hopkins University Press, 1991), p. 155.

47. *Skinner v. Oklahoma,* 316 U.S. 535 (1942).

48. *Eisenstadt v. Baird,* 405 U.S. 438, 453 (1972). For a rigorous philosophic defense of the right to reproduce, see John Robertson, "Procreative Liberty and the Control of Contraception, Pregnancy and Childbirth," *Virginia Law Review* 69 (1983): 405–62. And *idem,* "Embryos, Families, and Procreative Liberty: The Legal Structures of the New Reproduction," *Southern California Law Review* 59 (1986): 942–1041. For a thoughtful criticism of his position, see Maura A. Ryan, "The Argument for Unlimited Procreative Liberty: A Feminist Critique," *Hastings Center Report,* July–August 1990, pp. 6–12.

49. For an excellent discussion of this development, see Ronald Bayer, *Private Acts, Social Consequences: AIDS and the Politics of Public Health* (New Brunswick, N.J.: Rutgers University Press, 1989), esp. chapter 1.

50. Elizabeth Fox-Genovese, *Feminism without Illusions* (Chapel Hill: University of North Carolina Press, 1991), esp. chapters 2, 3.

51. Dorothy C. Wertz and John C. Fletcher, "Fatal Knowledge? Prenatal Diagnosis and Sex Selection," *Hastings Center Report,* May–June 1989, p. 24. For an opposed view, see Ruth Schwartz Cowan, "Genetic Technol-

ogy and Reproductive Choice: An Ethics for Autonomy," in Daniel J. Kevles and Leroy Hood, eds., *The Code of Codes: Scientific and Social Issues in the Human Genome Project* (Cambridge, Mass.: Harvard University Press, 1992), pp. 318–52.

52. Beth A. Fine, "The Evolution of Non-Directiveness in Genetic Counseling and Implications for the Human Genome Project," in Dianne M. Bartels *et al.*, eds., *Prescribing Our Future: Challenges in Genetic Counseling* (Hawthorne, N.Y.: Albine de Gryter, forthcoming 1993).

53. Carl R. Rogers, *Client-Centered Therapy: Its Current Practice, Implications, and Theory* (Boston: Houghton Mifflin, 1951), as cited in Fine, "The Evolution of Non-Directiveness in Genetic Counseling," p. 13.

54. Robert Nozick, *Anarchy, State, and Utopia* (New York: Basic Books, 1974), p. 315.

55. Wright, "Achilles' Helix," p. 27.

56. There is no direct evidence that children with genetic disabilities are being born unequally to the poor. However, access to genetic services (like other medical services) is strongly class-based. Thus, nearly a quarter of all children are born to women who received no prenatal health care at all. Specific studies of amniocentesis utilization rates demonstrate large geographic, racial, and socioeconomic differentials. Thus, David Sokal and colleagues found substantial geographic and racial variation among Georgia women aged forty and older, ranging from a sixty percent use rate among whites in the Atlanta and Augusta health districts to 0.5 percent among rural blacks. See David C. Sokal *et al.*, "Prenatal Chromosomal Diagnosis: Racial and Geographic Variation for Older Women in Georgia," *Journal of the American Medical Association* 244 (1980): 1355–57. Abby Lippmann summarizes the literature on socioeconomic differentials in "Prenatal Genetic Testing and Screening."

57. John Stuart Mill, "On Liberty," *Essential Works of John Stuart Mill* (New York: Bantam, 1855, rpt. 1961.)

58. *Ibid.*, pp. 353–54.

59. Mill also wrote that "every one has a right to live . . . But no one has a right to *bring* creatures into life, to be supported by other people. . . . There are an abundance of writers and public speakers, including many of most ostentatious pretensions to high feeling, whose views of life are so truly brutish that they see hardship in preventing paupers from breeding hereditary paupers in the workhouse itself. Posterity will one day ask, with astonishment, what sort of people it could be among whom such preachers could find proselytes." See John Stuart Mill, *Principles of Political Economy* (New York: Augustus M. Kelley, 1969; reprint of 1909 edition), p. 364. First published in 1848.

60. John Dunn, *Western Political Theory in the Face of the Future* (Cambridge: Cambridge University Press, 1979), p. 51.

Diane B. Paul and Hamish G. Spencer

7

DID EUGENICS REST ON
AN ELEMENTARY MISTAKE?

On the evidence of many genetics texts, of books on biology and society, and of histories of science, eugenicists were guilty of an astoundingly simple mistake. According to conventional accounts, which vary only in details, eugenic enthusiasts thought they could eliminate mental deficiency by segregating or sterilizing affected individuals. But a basic understanding of the Hardy-Weinberg principle suffices to destroy that illusion.

Eugenicists in the 1910s and 1920s attributed most mental defects to a recessive Mendelian factor (or in today's parlance, allele). But it is evident from the simple equation $p^2 + 2pq + q^2 = 1$ that if a trait is rare, the vast majority of deleterious genes will be hidden in apparently normal carriers. Selection against the affected themselves will thus be ineffectual. For example, even if all the affected were prevented from breeding, in a single generation the incidence of a trait at an initial frequency of 0.000100 would be reduced to just 0.000098 (and the allele frequency from 0.0100 to 0.0099). To reduce the incidence to half its original value (i.e., 0.000050) would require some forty-one generations, or about 1000 years. Tables in numerous genetics textbooks serve to make the point that hundreds of generations are required before a rare deleterious trait would disappear. Since a human generation lasts about twenty-five years, eugenical selection would be futile over any meaningful period. P. B. and J. S. Medawar express a common view: the eugenicists were ignorant and muddled (as well as foolish and inhumane). "Only a minority of the offending genes are locked up in the mentally deficient themselves," they explain, "so sterilizing them would not be effective."[1]

That selection is slow when genes are rare is not a new insight. Indeed, it originated with the Harvard geneticist Edward Murray East in 1917. In the same year, R. C. Punnett of Cambridge University refined the argument; Punnett's version was later popularized by J. B. S. Haldane in Britain and H. S. Jennings in the United States. "To merely cancel the deficient individuals themselves—those actually feebleminded—makes almost no progress toward getting rid of feeblemindedness for later generations," wrote the latter.[2] If the futility of sterilization and segregation were exposed so early and often, it might seem that the numerous geneticists who endorsed these policies were a remarkably dim-witted lot.

Whatever their personal and political failings, this explanation is implausible. R. A. Fisher was a social reactionary, as well as ardent eugenicist. But his worst enemies did not think him stupid. He unquestionably understood the implications of Hardy-Weinberg. Moreover, when Punnett first articulated these implications, he did so in an effort to expand eugenics' scope, not demonstrate its futility. Indeed, in the 1920s and 1930s, nearly all geneticists, including those traditionally characterized as opponents of eugenics, took it for granted that "mental defectives" should be prevented from breeding. To see why few geneticists of that period drew the conclusions that seem so obvious to their present-day successors, let us review the original arguments about the threat represented by carriers.

The "Real Menace" of the Feebleminded

In his 1917 essay "Hidden Feeblemindedness," East argued that neither the character nor scope of the problem of mental defect had been fully appreciated. While lauding efforts to cut off the stream of "defective germplasm" through segregation or sterilization of the affected, he thought the primary danger lay elsewhere, in the vast mass of invisible heterozygotes.

East had been influenced in this view by the American psychologist Henry H. Goddard, author of *The Kallikak Family: A Study in the Heredity of Feeblemindedness* (1912), an impressionistic study of a "degenerate" rural clan, and *Feeble-Mindedness: Its Causes and Consequences* (1914), a much longer work that discussed the meaning of the data for theories of inheritance. In the latter book, Goddard had argued that "normal-mindedness is dominant and is transmitted in accordance with the Mendelian law of inheritance."[3] His views were widely accepted. Thus Punnett could write in 1925 that no one "who has studied the numerous pedigrees collected by Goddard and others [could] fail to draw the conclusion that this mental state behaves as a simple recessive to the normal."[4] William E. Castle also praised Goddard's research and uncritically reported his results. "Goddard's evidence," he wrote in an influential textbook, "indicates that feeble-mindedness is a recessive unit-character."[5] As late as 1930, Jennings was able to assert that feeblemindedness was "the clearest case" of a recessive single gene defect.[6] While Paul Popenoe and R. H. Johnson did criticize Goddard's assumption that feeblemindedness resulted from a single gene, they accepted his claims that at least two-thirds of those affected owed their condition directly to heredity and that they numbered at least 300,000.[7]

Biometricians such as David Heron of the Galton Laboratory in London disparaged both the methods and logic used to reach this conclusion. In a passionate response to the stream of publications coming out of Charles B. Davenport's Eugenics Record Office in Cold Spring Harbor, Heron attacked almost every aspect of the Americans' work. Although his essay predated publication of *Feeblemindedness*, Heron's critique was as applicable to Goddard as Davenport. He concluded "that the material has been collected in an unsatisfactory manner, that the data have been tabled in a most slipshod fashion, and that the Mendelian conclusions drawn have no justification whatsoever."[8] Heron and the other biometricians were themselves ardent eugenicists, with "the highest hopes for the new science."[9] But they feared that eugenics would be crippled at birth by the American Mendelians' crude errors. Perhaps because of their unremitting anti-Mendelian rhetoric and personal style of attack, the biometricians' critiques were largely ignored by Mendelian geneticists on both sides of the Atlantic.

Davenport was one of the few *Mendelian* geneticists to criticize the category of feeblemindedness, which he characterized as a "lumber room" of different (and separately inherited) mental deficiencies.[10] He also noted the illogic of expecting a socially defined trait—a feebleminded person was considered "incapable of performing his duties as a member of society in the position of life to which he is born"—to be inherited as a simple Mendelian recessive.[11] But these were minor quarrels. Until the mid-1930s, Thomas Hunt Morgan was the only Mendelian geneticist to consistently repudiate Goddard's claim that social deviance was largely due to bad heredity.[12] In the 1925 edition of *Evolution and Genetics,* Morgan argued that much of the behavior tagged with that label was probably due to "demoralizing social conditions" rather than to heredity.[13] But Morgan's critique, like Heron's, had little impact.

East was thus one of many geneticists to conclude that feeblemindedness was genetic and transmitted as a Mendelian recessive. But he was the first to see the implications for eugenics. Even without benefit of the Hardy-Weinberg theorem, East understood that the number of apparently normal carriers must be much larger than those affected. In 1912, Davenport could offer the following advice:

Prevent the feebleminded, drunkards, paupers, sex-offenders, and criminalistic from marrying their like or cousins or any person belonging to a neuropathic strain. Practically it might be well to segregate such persons during the reproductive period for one generation. Then the crop of defectives will be reduced to practically nothing.[14]

Two years later, a committee of the American Breeders Association concluded almost as optimistically that two generations of segregation and sterilization would largely "eliminate from the race the source of supply of the great anti-social human varieties."[15] East realized that these predictions were wrong. The "real menace" of the feebleminded lay in the huge heterozygotic reserve, which constituted about seven percent of the American population, or one in every fourteen individuals. He warned: "Our modern Red Cross Knights have glimpsed but the face of the dragon."[16]

East's point was echoed by Punnett, who earlier had suggested that feeblemindedness could be brought under immediate control. Like many other geneticists, he felt "there is every reason to expect that a policy of strict segregation would rapidly bring about the elimination of this character."[17] But as a consequence of work for his influential 1915 book, *Mimicry in Butterflies,*[18] he changed his mind.

For his mimicry work, Punnett needed to know how fast a Mendelian factor would spread through a population.[19] He turned to his Cambridge mathematics colleague, H. T. J. Norton for help. Norton prepared a table (which appears as an appendix to Punnett's book) displaying the number of generations required to change the frequency of completely dominant or recessive factors at different selection intensities.[20] From the table, Punnett learned that selection could act with both surprising speed and, when the recessive factor was rare, extreme slowness. Two years after the *Mimicry* book appeared, Punnett called attention to the implications of the latter point for eugenics. Policies aimed at the affected, he argued, would take a distressingly long time to work. He employed a relatively well-understood condition to illustrate the point:

> Albinism, for example, behaves on whole as a recessive. Nevertheless, albinos appear among the offspring in an appreciable proportion of matings where either one or both parents are normal, and where no consanguinity can be detected. The same is true of feeblemindedness. This becomes less difficult to understand when we realize that the heterozygotes are bound greatly to outnumber the recessives whenever these form a small proportion of a stable population.[21]

While that argument had already been made by East, Punnett was able to work out its implications with much greater precision. Applying the Hardy-Weinberg formula, he concluded that more than ten percent of the population carried the gene for feeblemindedness. With G. H. Hardy's help, he also estimated the rate at which a population could be freed from mental defect by a policy of segregating or sterilizing the affected. He found the results depressing. Even under the unrealistic assumption that all the feebleminded could be prevented from breeding, their proportion in the population would only decline from:

1 in 100 to 1 in 1,000 in 22 generations
1 in 1,000 to 1 in 10,000 in 68 generations
1 in 10,000 to 1 in 100,000 in 216 generations
1 in 100,000 to 1 in 1,000,000 in 684 generations.

In other words, given Goddard's (unchallenged) estimate that about three in every thousand Americans were feebleminded as a result of genetic defect, it would take more than 8,000 years before their numbers were reduced to 1 in 100,000. Punnett thus concluded that eugenic segregation did not, contrary to his initial belief, offer a hopeful prospect.

Punnett, who served with Fisher on the Council of the Cambridge University Eugenics Society, did not intend to provide an argument against eugenics.[22] Like East, he concluded that if "that most desirable goal of a world rid of the feeble-minded is to be reached in a reasonable time, some method other than that of the elimination of the feeble-minded themselves must eventually be found."[23] That method would take advantage of the phenomenon of partial dominance. East had noted that complete dominance was rare among the characters studied by plant and animal breeders. He speculated that intelligence tests (which Goddard had introduced to America in 1908) could be used to identify heterozygotes, who would likely exhibit a lower mentality than the "pure normals." Punnett took up the suggestion, concluding his paper with a call for research to focus on carriers of defective genes.

Whatever his intention, Punnett's claim about the inefficacy of selection was seized on by critics of eugenic segregation and sterilization. For example, in 1923 the Central Association for Mental Welfare issued a pamphlet opposing sterilization, which it asserted "would have only a very limited effect in preventing mental deficiency."[24] In the same year, the Section of Medical Sociology of the British Medical Association sponsored a discussion on the issue of sterilizing mental defectives. The opponents of such a policy were clearly familiar with Punnett's argument. Thus, Dr. Joseph Prideaux, the mental and neurological inspector of the Ministry of Prisons, argued that if the proportion of mental defectives in the population were 3 or 4 per 1,000, it would be necessary to sterilize "some 10 per cent of the population, who were carriers of mental defect" (a policy he thought absurd) and that, moreover, "no really good result would be forthcoming until a very long period had elapsed."[25] Dr. H. B. Brackenbury, the Section's president, ended his summary of the discussion by remarking that the more the hereditary impact of rigorous segregation "was looked into the more certain aspects of it appeared to be disappointing," and noting that it had formerly been hoped that complete segregation or sterilization would rapidly eliminate the mentally defective population "but this was not so."[26]

R. A. Fisher realized that Punnett's calculations were misleading and easily employed to subvert the eugenic goals that he and Punnett shared. If the goal were to rid the world of the last few mental defectives, Fisher noted, the fact that thousands of generations are required to reduce their number to one in a billion would be meaningful. But if the calculations were extended to this point, "the reader would perhaps see the catch, and recognize that it would not matter if it took ten thousand generations to rid the world of its last lone feebleminded individual!"[27] Even on Punnett's unrealistic assumptions of a single gene for mental defect and random mating, Fisher argued, substantial progress could be achieved in the first few generations if affected individuals were prevented from breeding. Expressing the frequency of the defectives as so many per 10,000 easily demonstrates the point:

> From 100 to 82.6 in 1 generation
> From 82.6 to 69.4 in 1 generation
> From 69.4 to 59.2 in 1 generation.

Hence in the first generation alone, selection could remove more than seventeen percent of the affected persons.

Fisher's estimate is derived from Hardy's table, which represented an abstract calculation of the effects of selection, given assumptions about the initial number of affected. But the starting figures were chosen for ease of presentation rather than their assumed fit with reality. A standard estimate—and the one used by Punnett—was that three in a thousand individuals were feebleminded. Punnett's table could have been even more dramatic had he the skill to recalculate Hardy's numbers based on the lower initial frequency. But he understood little math. (In his 1916 referee's report on Fisher's classic paper, "The Correlation between Relatives on the Supposition of Mendelian Inheritance," Punnett wrote that it was of little interest to biologists but added: "frankly I do not follow it owing to my ignorance of mathematics."[28] If Fisher had used Punnett's estimate of the frequency of mental defect, the reduction in the first generation would have been about ten percent.[29]

Fisher also examined the effects of relaxing Punnett's assumption of random mating. This time, however, the result was more favorable to eugenics. Fisher assumed that the feebleminded constituted a larger proportion (one-sixteenth) of a smaller subsection (five percent) of the population, whose members mated only with others in that subsection. Hence, he incorporated a form of assortative mating into the model. While it seems reasonable to assume that the feebleminded would tend to mate among themselves, the five percent figure dramatically decreases the frequency of carriers, thus increasing the efficacy of selection. Even starting from the standard frequency of thirty affecteds per ten thousand people, Fisher calculated that mental defect could be reduced by thirty-six per cent in one generation.[30] Nevertheless, Fisher had shown for the first time that any form of assortative mating helped the eugenic cause.

Fisher's argument is often treated dismissively.[31] But Fisher diverged from Punnett and Jennings only in claiming that the affected tended to mate with each other (which would increase the frequency of homozygotes and thus speed selection) and that the trait was multifactorial. Both claims were eminently reasonable and at least as defensible as those of Punnett or Jennings—the conventional heroes of this historical fable—who were also more alarmist than Fisher. It was the progressive Jennings who asserted that "a defective gene—such a thing as produces diabetes, cretinism, feeblemindedness—is a frightful thing; it is the embodiment, the material realization of a demon of evil; a living self-perpetuating creature, invisible, impalpable, that blasts the human being in bud or leaf. Such a thing must be stopped wherever it is recognized."[32]

Fisher's primary criticism was levelled at the use of Hardy's table to demonstrate the inefficacy of selection. He was surely right in claiming that it was deceptive. What mattered to most eugenicists was the potential progress of selection in the next few generations. Here, Fisher demonstrated that eugenical policies could make a substantial difference. Even on Punnett's assumption of random mating, a substantial reduction in a single generation was possible.

In fact, all the geneticists agreed that the incidence of mental defect could be reduced by about ten percent in the first generation (and on the same reasoning, nineteen and twenty-six percent by the second and third generations respectively). Even Haldane, who regarded compulsory sterilization "as a piece of crude Americanism," thought it "would probably cut down the supply of mental defectives in the next generation by something of the order of 10 percent."[33] If some degree of assortative mating is assumed, the estimates would of course be higher. According to Jennings, the ostensible critic: "A reduction in the number of feebleminded by eleven percent [on the assumption of random mating], or still more, a reduction by thirty or forty percent [if mating is assortative], would be a very great achievement. And it could be brought about in no other way than by stopping propagation of the feeble-minded persons."[34]

Why are the estimates so high? It is often said that eugenics was based on a mistake about the efficacy of selection against rare genes. But this was not the eugenicists' error. The crucial point is that feeblemindedness was not considered rare, at least in comparison with a trait like albinism. Thus, Davenport wrote that eugenics was prompted by recognition of the "great proportional increase in feeble-mindedness in its protean forms—a great spread of animalistic traits—and of insanity."[35] Indeed, the *raison d'être* of the eugenics movement was the perceived threat of swamping by a large class of mental defectives. Numerous British and American studies and an increase in the institutionalized population seemed to indicate that the problem was rapidly worsening. In America it was commonly believed that from 300,000 to 1,000,000 persons were feebleminded as a result of genetic defect.[36] Those figures tended to increase as mental tests came into wider use to evaluate students, prisoners, inmates of poorhouses and training schools, immigrants at Ellis Island, and army draftees. In 1912, Goddard tested New York City schoolchildren and estimated that two percent were probably feebleminded.[37] The results of tests administered to army recruits during the World War I were even more alarming for they indicated that nearly half (47.3 percent) of the white draft—and eighty-nine percent of the black—was feebleminded.[38] Moreover, it seemed that the feebleminded were particularly prolific. For example, the British Royal Commission on the Care and Control of the Feeble-Minded reported in 1908 that defectives averaged seven children, normal couples only four; many other studies came to similar conclusions.[39]

Contemporary examples of the futility of eugenics often mention Tay-Sachs disease, phenylketonuria (PKU), or albinism. Selection against such diseases is certainly futile. But these textbook examples are almost invariably rare conditions whose effects are either lethal or minor. Both their frequency and consequences ensure that they would be of little interest to a eugenicist. Individuals with Tay-Sachs and (with a few exceptions) untreated PKU do not leave offspring. Albinos and treated phenylketonurics do reproduce, but these conditions are not disabling. The frequent employment of albinism in texts is probably an unconscious inheritance from Punnett's original article. In 1917, Punnett had few examples to choose among.

Applied to the historical eugenics movement, the argument about the futility of selection against rare genes is simply irrelevant. Given widely shared assumptions about the causes and incidence of mental defect, eugenic policies could be expected to substantially reduce the number of affected. In any case, geneticists in the 1920s would generally have favored such policies whatever their exact effect. In *Heredity and Eugenics,* Ruggles Gates summarized Punnett's argument, concluding that even if all mental defectives were prevented from reproducing "the most difficult part of the process of eliminating feeblemindedness from the germ plasm of the population would scarcely have begun."[40] But he ends the same chapter with a call for "the prevention of reproduction on the part of undesirables, such as the feebleminded," reasoning that, "such measures are necessary, not so much for the improvement of the race, as for arresting its rapid deterioration through the multiplication of the unfit."[41] Indeed, most geneticists would have assented to Jennings's claim that "to stop the propagation of the feebleminded, by thoroughly effective measures, is a procedure for the welfare of future generations that should be supported by all enlightened persons. Even though it may get rid of but a small proportion of the defective genes, every case saved is a gain, is worth while in itself."[42]

Like Jennings, Lancelot Hogben is often portrayed as an opponent of eugenics. He did criticize some advocates of sterilization for exaggerating the urgency of the problem and the results they could achieve—fearing that overstatement would harm the cause. He also invoked Fisher to argue that there is no need to overstate potential results. That we cannot do everything "is not a valid reason for neglecting to do what little can be done."[43] That point was echoed by Edwin G. Conklin, who like Jennings and Hogben, criticized some eugenic proposals. Conklin once remarked that sterilizing all the inmates of public institutions was "like burning down a house to get rid of the rats."[44] But he did not oppose sterilization of the feebleminded. On the contrary, he asserted that "all modern geneticists approve the segregation or sterilization of those who are known to have serious hereditary defects, such as hereditary feeble-mindedness, insanity, etc." Conklin asked of the American Eugenics Society's proposed sterilization policy: "Can any serious objection be urged to such a law?"[45] In 1930, this question was unambiguously rhetorical.

Nearly all geneticists of the 1920s and 1930s—including those traditionally characterized as opponents of eugenics—took for granted that the "feebleminded" should be prevented from breeding. Moreover, nearly everyone agreed on the scientific facts. Punnett, East, Fisher, Jennings, and even Haldane made roughly the same estimates as to the speed and scope of eugenical selection. But in respect to social policy, the facts did not speak for themselves. They required interpretation in light of other assumptions and goals. Thus, Haldane opposed sterilization, arguing that "with mental defects as with physical defects, if you once deem it desirable to sterilize I think it is a little difficult to know where you are to stop."[46] That is a powerful argument. But it is a social, not a scientific one. Lionel Penrose was an even more vehement and consistent critic of eugenics. An expert in the genetics of mental deficiency, he stressed the heterogeneity of its causes and the modest influence of eugenic measures in reducing its incidence. But his main argument was ethical. Penrose maintained that the best index of a society's health is its willingness to provide adequate care for those unable to care for themselves.[47]

The Hardy-Weinberg theorem meant different things to different people. To those already disposed against eugenics, it proved that policies to prevent the feebleminded from breeding were not worth the effort. There is no reason that those disposed in favor of eugenics should draw the same conclusion. Whether a ten percent reduction in incidence is large or small is not a question science can answer. Indeed, one may concede that the percentage reduction is very small yet consider it worthwhile. Thus, at the close of a long discussion of the implications of Hardy-Weinberg, Curt Stern remarked: "To state that reproductive selection against severe physical and mental abnormalities will reduce the number of affected from one generation to the next by only a few per cent does not alter the fact that these few per cent may mean tens of thousands of unfortunate individuals who, if never born, will be saved untold sorrow."[48] A similar point was made by the Swedish Commission on Population in its 1936 special report on sterilization. After acknowledging the falsity of the earlier belief that sterilization would result in a rapid improvement of the population, the authors note that it would still result in gradual improvement while preventing possible deterioration and that "whenever an eugenic sterilization is carried out . . . in the specific case the operation will prevent the birth of sick or inferior children or descendants. Owing to this, sterilization of hereditary sick or inferior human beings is still a justified measure, beneficial to the individual as well as to society."[49] Thus, it may not matter if the reduction in absolute numbers is minuscule. Indeed, the rate of selection is simply beside the point if one assumes with Jennings that "the prevention of propagation of even one congenitally defective individual puts a period to at least one line of operation of this devil. To fail to do at least so much would be a crime."[50]

We began by asking whether eugenics rested on an elementary mistake. To the extent that support for eugenical segregation and sterilization was based on the assumption that "it would be possible at one fell stroke [to] cut off practically all of the cacogenic varieties of the race,"[51] a loose definition of "feeblemindedness," as well as acceptance of Goddard's shoddy data and defective logic, the answer is yes. But it was possible to recognize these flaws and still remain a eugenicist, as the example of David Heron demonstrates. Moreover, what is usually characterized as the eugenicists' most obvious error—a failure to understand the implications of the Hardy-Weinberg theorem—was a mistake few geneticists made after 1917. By the 1920s, they well understood that the bulk of genes for mental defect would be hidden in apparently normal carriers. For most geneticists, this appeared a better reason to widen eugenic efforts than to abandon them.

It is often said that support for eugenics declined in the 1930s as its scientific errors were exposed. But the eugenics movement grew stronger during the Depression.[52] In the United States, the number of sterilizations climbed. The procedure was legalized in Germany (1933), the Canadian province of British Columbia (1933), Norway (1934), Sweden (1934), Finland (1935), Estonia (1936), and Iceland (1938). Denmark, which in 1929 had legalized "voluntary" sterilization, permitted its coercive use on mental defectives in 1934. These laws were generally applauded by geneticists.

In 1918, Popenoe and Johnson wrote that "so few people would now contend that two feeble-minded or epileptic persons have any 'right' to marry and perpetuate their kind, that it is hardly worth while to argue the point."[53] Assumptions we now take for granted they thought too absurd even to require challenging. The inversion of these assumptions in recent decades is best explained by political developments. Revelations of Nazi atrocities, the trend toward respect for patients' rights in medicine, and the rise of feminism have converged to make reproductive autonomy a dominant value in our culture. In 1914, a committee of the American Breeders Association asserted that "society must look upon germ-plasm as belonging to society and not solely to the individual who carries it."[54] Few today would profess such a view. A change in values, and not the progress of science, explains why contemporary Swedes would be unlikely to concur with the 1936 commission that criticized as "extremely individualistic" the concept that individuals have a right to control their own bodies.[55]

It is not our superior quantitative skills that explain why we today draw very different implications from the Hardy-Weinberg theorem. There was nothing wrong with most eugenicists' math. Our concept of rights, however, is much more expansive than theirs. And that is why the same equation holds different lessons for them than it does for us.

Notes

1. P. B. Medawar and J. S. Medawar, *The Life Sciences: Current Ideas in Biology* (New York: Harper and Row, 1977), p. 60.

2. H. S. Jennings, "Health Progress and Race Progress: Are They Incompatible?" *Journal of Heredity* 18 (1927): 273; J. B. S. Haldane, *The Inequality of Man and Other Essays* (London: Chatto and Windus, 1928, rpt. 1932), p. 105.

3. H. H. Goddard, *Feeblemindedness: Its Causes and Consequences* (New York: Macmillan, 1914), p. 556.

4. R. C. Punnett, "As a Biologist Sees It," *The Nineteenth Century* 97 (1925): 704.

5. W. E. Castle, *Genetics and Eugenics* (Cambridge, Mass.: Harvard University Press, 1927), p. 355.

6. H. S. Jennings, *The Biological Basis of Human Nature* (London: Faber and Faber, 1930), p. 238.

7. P. Popenoe and R. H. Johnson, *Applied Eugenics* (New York: Macmillan, 1918), pp. 105, 175.

8. D. Heron, "Mendalism and the Problem of Mental Defect. I. A Criticism of Recent American Work," in *Questions of the Day and of the Fray No. VII.* London: Cambridge University Press, 1913), p. 61.

9. *Ibid.*, p. 4.

10. C. B. Davenport, "The Inheritance of Physical and Mental Traits of Man and Their Application to Eugenics," in W. E. Castle, J. M. Coulter, C. B. Davenport, E. M. East, and W. L. Porter, eds., *Hereditary and Eugenics* (Chicago: University of Chicago Press, 1912), pp. 269–88.

11. P. Popenoe, "Feeblemindedness," *Journal of Heredity* 6 (1915): 32. See also Davenport, "Inheritance," p. 236; Samuel J. Holmes, *Studies in Evolution and Eugenics* (New York: Harcourt, Brace, 1923), pp. 121–23; Albert E. Wiggam, *Fruit of the Family Tree* (Indianapolis: Bobbs-Merrill, 1924), pp. 56–58.

12. D. Barker, "The Biology of Stupidity: Genetics, Eugenics, and Mental Deficiency in the Inter-war Years," *British Journal of the History of Science* 22 (1989): 347–75.

13. T. H. Morgan, *Evolution and Genetics,* 2d ed. (Princeton: Princeton University Press, 1925), p. 201.

14. Davenport, "Inheritance," p. 286.

15. H. H. Laughlin, "Report of the Committee to Study and to Report on the Best Practical Means of Cutting off the Defective Germ-Plasm in the American Population. I. The Scope of the Committee's Work." Eugenics Record Office, Bulletin No. 10A. Cold Spring Harbor, N.Y., 1914, p. 60.

16. E. M. East, "Hidden Feeblemindedness," *Journal of Heredity* 8 (1917): 215.

17. R. C. Punnett, "Genetics and Eugenics," in *Problems in Eugenics: Papers Communicated to the First International Eugenics Congress* (London: Eugenics Education Society, 1912), p. 137.

18. R. C. Punnett, *Mimicry in Butterflies* (Cambridge: Cambridge University Press, 1915).

19. W. B. Provine, *The Origins of Theoretical Population Genetics* (Chicago: University of Chicago Press, 1971), p. 137; J. H. Bennett, *Introduction to Natural Selection, Heredity, and Eugenics* (Oxford: Clarendon Press, 1983), pp. 8–10.

20. Punnett, *Mimicry in Butterflies*.

21. Punnett, "Eliminating Feeblemindedness," p. 465. Indeed, most texts continue to treat albinism—as Punnett did—as a single-locus defect. D. S. Falconer, in *Introduction to Quantitative Genetics,* 3d ed. (Harlow, G.B.: Longman, 1989), is one of the few exceptions. But it has long been known that albinism arises from the actions of recessive genes from at least two loci (See V. A. McKusick, *Mendelian Inheritance in Man: Cata-*

logue of Autosomal Dominant, Autosomal Recessive, and X-Linted Pheno-types. 10th ed. [Baltimore: Johns Hopkins University Press, 1992]). Consequently, the incidence of homozygous recessives for a particular locus is lower than most texts suggest and eugenic selection against albinism would be even less efficacious.

22. Bennett, p. 12.

23. Punnett, "Eliminating Feeblemindedness," p. 464.

24. Central Association for Mental Welfare, "Sterilisation and Mental Deficiency," *Studies in Mental Inefficiency* 4 (July 15, 1923): 12.

25. "General Discussion: Mental Deficiency in Its Social Aspects," *British Medical Journal,* April 11, 1923, p. 231.

26. "President's Summary of the Discussion," *British Medical Journal,* April 11, 1923, pp. 233–34.

27. R. A. Fisher, "The Elimination of Mental Defect," *Eugenics Review* 26 (1924): 114. Reprinted in J. H. Bennett, ed., *Collected Papers of R. A. Fisher.* Vol. 1. (Adelaide, Aus.: University of Adelaide, 1971).

28. Bennett, *Introduction to Natural Selection,* p. 116, n. 12.

29. In emphasizing that his presentation was based on Punnett's assumptions, Fisher traded on this weakness. A reader could easily presume that Fisher employed assumptions favorable to Punnett's case. He did not. "In a single generation," Fisher wrote, "the load of public expenditure and personal misery caused by feeblemindedness would be reduced by over 17 per cent." If based on the figures in Punnett's table, this estimate is correct but also misleading. Fisher did add that if the starting point had been thirty instead of a hundred (per ten thousand), the reduction in one generation would be "over 11 per cent" (p. 114). But he failed to note that this is the relevant figure. In fact, Fisher's 11 per cent figure is still inflated. The true reduction is approximately 10.1 percent. Haldane discreetly gives the correct value—"something of the order of 10 percent"— but less mathematically inclined writers such as Jennings appear not to have noticed Fisher's error. See J. B. S. Haldane, *Heredity and Politics* (London: Unwin, 1938), p. 88; H. S. Jennings, *The Biological Basis of Human Nature* (New York: W. W. Norton, 1930), p. 242.

30. Fisher, p. 115.

31. See Daniel J. Kevles, *In the Name of Eugenics: Genetics and the Uses of Human Heredity* (New York: Knopf, 1985), p. 165; D. Barker, "The Biology of Stupidity: Genetics, Eugenics and Mental Deficiency in the Inter-War Years," *British Journal of the History of Science* 22 (1989): 347–75.

32. H. S. Jennings, "Health Progress," p. 274.

33. Haldane, *Heredity and Politics,* pp. 80, 88.

34. Jennings, *The Biological Basis of Human Nature,* p. 242.

35. Davenport, p. 308.

36. Popenoe and Johnson, p. 158.

37. Goddard, "Ungraded Classes." Report on Educational Aspects of the Public School Systems of the City of New York, Part II, Subdivision I, Section E. City of New York, 1911–12.

38. Robert M. Yerkes, ed., *Psychological Examining in the United States Army*. Vol. 15. *Memoirs of the National Academy of Sciences*. Washington, D.C., 1921.

39. Diane B. Paul, *Controlling Human Heredity: 1865 to the Present* (Atlantic Highlands: Humanities Press, 1995), pp. 62, 78.

40. R. R. Gates, *Heredity and Eugenics* (London: Constable, 1923), p. 159.

41. *Ibid.*, p. 251.

42. Jennings, *The Biological Basis of Human Nature*, p. 238.

43. Lancelot Hogben, *Genetic Principles in Medicine and Social Science* (London: Williams and Norgate, 1931), p. 207.

44. E. G. Conklin, *Heredity and Environment in the Development of Man*, 2d ed. (Princeton: Princeton University Press, 1916), p. 438.

45. E. G. Conklin, "The Purposive Improvement of the Human Race," in E. V. Cowdry, ed., *Human Biology and Population Improvement*, (New York: Hoeber, 1930), pp. 577–78.

46. Haldane, *Heredity and Politics*, p. 89.

47. Lionel S. Penrose, *Biology of Mental Defect* (London: Sidgwick and Jackson, 1949). See also Kevles, especially pp. 151–63.

48. C. Stern, *Principles of Human Genetics* (San Francisco: Freeman, 1949), p. 538.

49. Quoted in G. Broberg and M. Tydén, "Eugenics in Sweden: Efficient Care," in G. Broberg and N. Roll-Hansen, eds., *Eugenics and the Welfare State: Sterilization Policy in Denmark, Sweden, Norway, and Finland* (East Lansing: Michigan State University Press, 1996), p. 106.

50. Jennings, "Health Progress," p. 274.

51. Laughlin, p. 47.

52. Paul, *Controlling Human Heredity*, pp. 72–90.

53. Popenoe and Johnson, p. 170.

54. Laughlin, p. 16.

55. Broberg and Tydén, p. 138.

8

Eugenic Origins of Medical Genetics

Introduction

Sheldon Reed coined the expression "genetic counseling" in 1947, the same year he succeeded Clarence P. Oliver as director of the University of Minnesota's Dight Institute for Human Genetics. In reflections written more than a quarter-century later, Reed noted that the term had occurred to him "as a kind of genetic social work without eugenic connotations." Sharply distinguishing the aims of eugenics and counseling, he explained that whereas the former promotes the interests of the larger society, the latter serves the interests of individual families—as families perceive them. Reed never denied that he or other postwar medical geneticists were concerned with population improvement. But he maintained that counseling served a different purpose. Commenting on the history of the Dight Institute, Reed asserted: "There were certainly no attempts to benefit society as a whole in dealing with these families. This was not thought of as a program of eugenics."[1]

The historical record suggests a rather more complex story. In the 1950s and 1960s, genetic counseling was characterized by most of its practitioners as an extension of eugenics. Thus, in a 1950 application to the Rockefeller Foundation, Reed himself stated: "Counseling in human genetics is the modern way of carrying on a program in Eugenics . . . the term 'Eugenics' has fallen by the wayside and 'Counseling in Human Genetics' is taking its place."[2] And two years later, he wrote that "it could be stated as a principle that the mentally sound will voluntarily carry out an eugenics program which is acceptable to society *if* counseling in genetics is available to them."[3]

Given the protean meanings of "eugenics," a plausible case can be made for each of his claims. In some respects, Reed seems decidedly anti-eugenicist. Long before neutrality became fashionable, Reed argued that counselors ought not to impose their own values on their clients. The role of the counselor, he consistently argued, was "to explain thoroughly what the genetic situation is but the decision must be a personal one between the husband and wife, and theirs alone."[4] Moreover, he wrote that the net effect of counseling might well be dysgenic. Reed often noted that the desire to compensate for the birth of an affected child was usually strong, while the recurrence risk was typically lower than the family had feared. Hence on balance, the effect of genetic counseling would be to encourage people to have more children than they otherwise would—thereby spreading the defective gene through offspring who were normal carriers.[5] Reed's views on the appropriate stance for counselors and probable impact of their work hardly seem consistent with his characterization of counseling as the modern form of eugenics.

But Reed also believed that normal people could be relied on to make "rational" decisions—that is, to avoid bearing children at high risk for seriously abnormal conditions. Thus, the *impact* of counseling could be described as eugenic even if its aim were relief of individual suffering rather than changes in the distribution of births and its means—provision of information to those who asked for it—were wholly voluntary. Second, while counseling might increase the incidence of particular disease genes, Reed and most of his peers assumed both that mental traits were more important than physical ones and that individuals who availed themselves of counseling services were generally impressive in intelligence and character. Therefore, counseling could be considered dysgenic in respect to disease and eugenic in respect to behavior. Third, any impact on the incidence of disease genes would be felt only in the distant future, with the immediate consequence a reduction in the birth of affected children. Whether counseling appears to be dysgenic or eugenic is thus also a function of whether concern focuses primarily on long- or short-term effects. Depending on which factors are emphasized and how they are interpreted, counseling could be equated with eugenics—or with its antithesis.

That is why Reed could plausibly claim in the 1950s that counseling was a form of eugenics and with equal plausibility in the 1970s that it was not. But the question remains: why would anyone *want* to identify counseling with such an ostensibly discredited enterprise? As we will see, "eugenics" in the 1950s still retained positive connotations for many scientists and their sponsors. Indeed, following Watson and Crick's 1953 discovery of the double-helical structure of DNA, it enjoyed a temporary resurgence in popularity. In the years immediately following the end of World War II, the word "eugenics" virtually vanished from scientific journals. While arguments for selective breeding did not disappear, they were now mostly relegated to the conclusions of more general articles on eugenics or birth control.[6] But publication of the Watson and Crick paper seems to have emboldened some geneticists. At the same time, concern about the genetic consequences of increased exposures to ionizing radiation was mounting. "Eugenics" began to reappear in the titles of articles in scientific journals. And as we will see, their arguments became more forceful and direct.

Thus, there was little reason to avoid the association of eugenics with counseling. By the mid-1970s, however, "eugenics" had once again become a term of abuse. The shift in emotional resonance was accompanied by a contraction in the meaning of the word. Now that the association was damning, "eugenics" was typically restricted to compulsory programs. On the narrow definition, it was made unambiguously distinct from counseling.

This essay explores the ethos of medical genetics and genetic counseling as these fields developed in America and Britain in the two decades following World War II. In prior decades, some individual geneticists, such as Charles Davenport at Cold Spring Harbor and Lawrence Snyder and Madge Macklin at Ohio State University, had provided "marriage advice" to those who sought their help. Formal clinics had been established in Germany and Denmark during the 1930s. But in the Anglo-American world, genetic counseling was first institutionalized in the 1940s, when clinics were opened in Britain at the Hospital for Sick Children and in the United States at the Universities of Michigan and Minnesota. This essay asks: What were the aims of the field in the two decades following the end of World War II? How were sometimes disparate goals reconciled? How were these goals reflected in clinical practice? Are some of the tensions that once marked the field still manifest and, if so, in what ways?

Establishing Medical Genetics: Scientists and Their Sponsors

The Dight Institute was founded in 1941 with the explicit aim of promoting eugenics.[7] A passage from the second annual Dight lecture exemplifies its ethos in the Institute's early years:

> In a commendable exhibition of sympathy and generosity, the nonproductive classes of society are being cared for on a plane of living which our productive members cannot afford for themselves. Very little is being done to protect our social system by our procedure in respect to these dysgenic classes. The burden has already become so great that a surprising amount of our public expenditures in so-called normal times goes for the care of these nonproductive classes.[8]

The eugenic orientation reflected the aims of Charles Fremont Dight, a Minneapolis physician who left his estate to the University of Minnesota "to promote biological race betterment." Dight's many causes included birth control, socialism, and eugenics. A president of the Minnesota Eugenics Society, member of the Minnesota Birth Control League, unsuccessful Congressional candidate of the Public Ownership Party (1906), and socialist alderman for Minneapolis's 12th ward (1914–1918), Dight lobbied for a state sterilization law and, after its passage in 1925, for its extension to the non-institutionalized.[9] The equally eccentric Charles M. Goethe, a bank president and founder of the playground movement, also left much of his estate to what was by then the Dight Institute for its eugenic work while the reactionary textile magnate "Colonel" Wycliffe C. Draper supported the Department of Medical Genetics at the Bowman-Gray School of Medicine in Winston-Salem, North Carolina (the first Department of medical genetics at an American university) and its director, C. Nash Herndon.[10] Draper wished to fund individuals and institutions with the proper attitudes toward "(a) miscegenation, (b) immigration quotas, (c) improving population quality by (1) positive, (2) negative measures."[11]

With the exception of the U.S. Public Health Service, which funded cancer-related projects, virtually all institutional patrons of work in medical genetics and genetic counseling also had eugenic motivations. These included the Rockefeller, Carnegie, Wenner-Grenn, McGregor, and Rackham foundations, the Commonwealth and Pioneer Funds (the last founded by Draper in 1937), and the American Eugenics Society.[12]

Across a wide political spectrum, scientists with an interest in medical genetics agreed that the field should serve to improve the race. To many scientists, it seemed self-evident that reproduction was properly a social and not just a private matter. Thus, Ashley Montagu asserted in 1959 that "there can be no question that infantile amaurotic family idiocy is a disorder that no one has a right to visit upon a small infant. Persons carrying this gene, if they marry, should never have children, and should, if they desire children, adopt them."[13] Linus Pauling, who believed that genetic defects were a primary source of human misery, proposed in 1968 that all young people be tested for the presence of the sickle-cell and other deleterious genes and a symbol be tattooed on the foreheads of those found to be carriers.[14] In his 1970 presidential address to the American Association for the Advancement of Science, Bentley Glass speculated on the changes that would be prompted by exponential population growth. He wrote: "In a world where each pair must be limited, on the average, to two offspring and no more, the right that must become paramount is not the right to procreate, but rather the right of each child to be born with a sound physical and mental constitution, based on a sound genotype. No parents will in that future time have a right to burden society with a malformed or a mentally incompetent child."[15]

Through the 1960s, most of the leading figures in medical genetics—including Oliver, Curt Stern, Lee R. Dice, Herluf Strandskov, Gordon Allen, William Allan, C. Nash Herndon, Franz Kallmann, and Harold Falls, Madge Macklin, and F. Clarke Fraser in the United States and Canada, Eliot Slater and Cedric Carter in Britain, and Tage Kemp in Denmark—bluntly described their work as a form of "eugenics." The links between medical genetics and eugenics are nicely illustrated by the early history of the American Society of Human Genetics (ASHG), which was founded in 1948. Four of the first five Presidents—Dice, Snyder, Oliver, and Kallmann—were members of the Board of the American Eugenics Society. (Herndon also served as President of the ASHG in 1955, Reed in 1956, Stern in 1957, and Macklin in 1958).

The exception was H. J. Muller. An ardent critic of "mainline" eugenics, Muller argued that eugenics in capitalist societies was hopelessly distorted by class and racial bias. But he was not opposed to eugenics per se. His 1949 Presidential Address to the American Society of Human Genetics, published as "Our Load of Mutations," argued that identifying individuals carrying more than their share of the genetic load and convincing them not to reproduce was a matter of urgent necessity. But he consciously avoided using the word "eugenics" to describe his scheme. In Muller's view, eugenic goals were best pursued under another rubric.[16] Thus, Muller differed from his peers in his view of appropriate tactics, not ultimate goals. He would certainly have agreed with Lee R. Dice, the first director of the University of Michigan's Heredity Clinic, that "the heredity of the population should be of at least as much concern to each commonwealth as infectious diseases."[17] Summarizing a 1952 panel discussion on genetic counseling, Dice asserted: "We must give due concern to the possibility of eliminating, or, perhaps, of perpetuating, undesirable or desirable genes. We must not only be concerned with the particular family concerned, but also with whether or not harmful heredity may be continued or spread in our population."[18] That this was the prevailing view explains why the practice of genetic counseling was usually directive, and sometimes strongly so.

In the 1950s, genetic services came to be centered in major medical centers, and physicians, who are trained to be directive, assumed a greater role. A common view—that the attitude of neutrality "originated with counselors who were not engaged in patient care and who may have felt some reluctance, therefore, to enter into the lives of their counselees in the way a practicing doctor frequently does"—may thus seem plausible.[19] But most of the research-oriented Ph.D. geneticists felt a similar responsibility to guide their clients. The views of Reed's predecessor, C. P. Oliver, were typical: "A geneticist should prevail upon some persons to have at least their share of children as well as show a black picture to those with the potentiality of producing children with undesirable traits."[20]

While the early postwar literature on counseling is replete with assertions that reproductive decisions belong to parents, they do not necessarily imply support for a neutral stance. Thus, Oliver declared that parents should make their own decisions after they have been given all the facts. But counselors would also "make the picture as dark as possible" to help particular parents reach the conclusion that it would be best not to have more children.[21]

Some clinicians expressed optimism that, advised of their hereditary defect, clients would generally follow their doctor's advice. But most thought they needed at least a gentle push. Thus, C. Nash Herndon, one of the two original staff members of the Heredity Clinic and first director of the Department of Medical Genetics at Bowman-Gray School of Medicine, thought that "the counselor should attempt to encourage the marriage of persons of average or superior physical or mental capabilities, and should encourage such persons to have families. On the other hand, those with obvious hereditary defects ... should be discouraged."[22] Franz Kallmann similarly believed that "persons requesting genetic advice cannot always be presumed to be capable of making a realistic decision as to the choice of a mate, or the advisability of parenthood, without support in the form of directive guidance and encouragement."[23] In his popular textbook, Curt Stern even anticipated the day when:

Natural selection will be superseded by socially decreed selection. In the course of time ... the control by man of his own biological evolution will become imperative, since the power which knowledge of human genetics will place in man's hands cannot but lead to action. Such evolutionary controls will be world wide in scope, since, by its nature, the evolution of man transcends the concept of unrestricted national sovereignty.[24]

Distinguishing Good Eugenics from Bad

On conventional accounts, eugenics was wounded by the Depression and died with revelations of Nazi atrocities following World War II. Word and concept are said to have fallen into disrepute. But this generalization is much too broad. It is true that much of the public soon came to equate eugenics with the policies of the Third Reich. It is also true that the 1950s witnessed developments in medical genetics—broadly defined—that had little if any connection to eugenics. Research on the "inborn errors of metabolism" first identified by Sir Archibald Garrod at the turn of the century is particularly important for it showed that some genetic diseases could be treated—a crucial step in the expansion of genetic services in the 1960s.[25]

The idea that a genetic disease might be treatable was first suggested to a broad audience by Lionel Penrose in his 1946 inaugural lecture as professor of eugenics and head of the Galton Laboratory at University College London. In "Phenylketonuria: A Problem in Eugenics," Penrose stressed the complex causes of mental deficiency and argued that eugenical measures could have only a slight impact on its incidence. He also suggested that phenylketonuria (or PKU), although a genetic disease, might one day be ameliorated through an environmental intervention.[26]

That day was in fact not far off. The severe mental retardation and other symptoms associated with PKU result from an excess of blood phenylalanine. (Due to a defective liver enzyme, phenylketonuric individuals are unable to catalyze the conversion of the essential amino acid phenylalanine to another amino acid, tyrosine). In the mid-1950s, a number of groups began experimental treatment of affected infants and children with low-phenylalanine diets. While their efforts initially met with only mixed success, the prospect of treating a genetic disease generated great excitement among public health officials, parents' groups, and the press.[27] In an influential 1958 report commissioned by the National Association of Retarded Children, the neurologist Richard Masland wrote: "The fact that a disease is hereditary does not indicate that there is no form of therapy conceivable or that sterilization or other eugenic practices are the only hopes for modification of the problem. The modification of the stressful features of our environment, in the broadest sense of the word, may be an entirely proper and effective means of dealing with many genetic disorders."[28]

Two years later, a cheap and simple blood test, suitable for mass screening, became available. Within a decade, newborn screening for PKU and other inborn errors had become routine in the United States, Britain, and much of Europe. Although all the metabolic disorders were rare, screening advocates successfully argued that the cost to the state of lifetime institutionalization for untreated individuals greatly exceeded the cost of the screening programs and diet required by affected infants and children.

Thus, a competing model for medical genetics had already emerged in late 1950s. While preserving the orientation toward public health, and its associated cost-benefit language, newborn screening problems focused on treatment, not selective breeding. (Most researchers in the area of human metabolic disorders were physicians and biochemists rather than geneticists). At the same time, work in human cytogenetics was rapidly expanding. Joe-Hin Tijo, Albert Levan, Charles Ford, Paul Polani, Murray Barr, and Jerome Lejune, among others, greatly refined the analysis of chromosomes and thus laid the scientific groundwork for prenatal diagnosis.[29] Like the metabolic researchers, the cytogeneticists were generally uninterested in eugenics. Thus, medical genetics had already begun to fragment as younger scientists with different interests entered the field.

At the same time, many geneticists whose professional careers began before World War II worried that eugenics' rational core would be abandoned in the reaction to past abuses. Some even thought that a program of artificial selection was made more urgent by postwar military and medical uses of radiation, which they assumed were increasing the human mutation rate. Throughout the 1940s, 1950s, and even 1960s, few geneticists objected to the characterization of applied medical genetics as "eugenics." (Lionel Penrose, who insisted that his position as professor of eugenics be retitled professor of human genetics, is a major exception; James Neel is another.) For example, Lawrence Snyder noted that the practical applications of a knowledge of genetics include "the setting up of eugenic programs for the protection and improvement of society."[30] In an article on X-linked mental retardation, William Allan, C. Nash Herndon, and Florence Dudley wrote that "when a sufficient body of data has been assembled to permit us to predict with accuracy the probable occurrence of mentally deficient children, we believe that a program of negative eugenics will do much to reduce the supply of disastrous children from these causes."[31] In the 1950s and early 1960s, leading figures in the field routinely defined medical genetics as a worthy form of eugenics.[32]

Thus, older geneticists generally continued to speak the language of eugenics, condemning past abuses but also taking for granted that reproduction was an act with social consequences and was thus legitimately a matter of social concern. The eugenics of the past, they conceded, was distorted by racial and class prejudice and simplistic scientific assumptions. But they insisted that eugenics has a rational core, which should be preserved. Some genes are unreservedly bad. Those that produce Tay-Sachs disease, muscular dystrophy, Huntington's Chorea, and other serious conditions bring only misery to their bearers and unnecessary expense to society. The struggle to eliminate disease genes must be sharply demarcated from past policies that targeted ethnic and religious minorities and the poor.[33]

That was also the position adopted by the American Eugenics Society. Under the leadership of Frederick Osborn, distinguished scientists such as Theodosius Dobzhansky and Tracy Sonneborn were recruited to its board of directors. Osborn also increasingly turned the Society's efforts toward the apparently neutral fields of birth control and human genetics. In 1954, the first issue of its new journal announced a series on "heredity counseling." During the next four years, an article on this theme appeared in almost every issue. In fact, between 1954 and 1958, it published more articles on counseling than any other topic. The contributors constituted a virtual "who's who" in the field, most stressing its eugenic potential.

Public aversion to anything labelled eugenics ultimately swamped the "reform eugenics" movement. The Society's general membership declined steeply. In a concession to public sentiment, its journal, the *Eugenics Quarterly,* was renamed *Social Biology* in 1968. Although some geneticists continued to employ the label into the 1970s, it was by then generally recognized that a successful eugenics program must be called something else. Commenting on the new title, Osborn remarked that "birth control and abortion are turning out to be great eugenic advances of our time. If they had been advanced for eugenic reasons, it would have retarded or stopped their acceptance."[34] Or, as he wrote in a popular 1968 book: "Eugenic goals are most likely to be attained under a name other than eugenics."[35]

The commitment to birth control is not surprising. In the 1910s and 1920s, eugenicists had divided on the question of its value. Some feared that the widespread practice of contraception "would prejudice the production of sufficient babies by the competent and far-seeing section of the community."[36] Others argued that the fittest members of society already limited their births and that the extension of contraception would therefore improve the race. Over time, as the futility of preventing its spread among middle- and upper-class women became increasingly evident, many eugenicists converted to the birth control cause.

Furthermore, changing public opinion had left eugenicists with few other options. Coercive programs were no longer in vogue. Birth control advocates argued that, at least in respect to the normal population, there was no need for compulsion. The race would be improved by the voluntary actions of poor women who wanted to limit their births. The American Eugenics Society began to aggressively promote contraception. In 1952, Osborn was appointed the first director of the Population Council, an organization funded by John D. Rockefeller III to promote what was now often called "family planning."

But what could genetic counseling actually offer eugenicists? The answer is: very little. Those who turned to geneticists for advice (typically parents who already had a child affected with a genetic disorder or who were anxious about transmitting a trait that ran in the family) were confronted with a stark choice based on often vague estimates of risk. Until abortion was legalized in the United States by the 1973 Supreme Court decision in *Roe v. Wade* and in Britain by a 1967 Act of Parliament, the only legal way to avoid genetic risk was not to reproduce. But the right to terminate a pregnancy would have had little impact in the absence of practical methods for detecting genetic disorders during pregnancy. In the 1960s the first such method—amniocentesis—was developed, and by the mid-1970s, it had become a routine part of clinical practice. The convergence of prenatal diagnosis and legalized abortion produced explosive growth in the field of genetic counseling. But in the three decades following World War II, it was practiced on too small a scale to make an appreciable difference in the population frequencies of the diseases in question. Thus, from a eugenical standpoint, counseling was insignificant. Moreover, its impact was as likely to increase as reduce the incidence of particular disease genes (which is why Reed thought that genetic counseling might well be dysgenic).

The Society's embrace of genetic counseling in part reflects its limited options in the postwar period. It also reflects the fact that eugenicists have in fact rarely focused on long-term effects. With few exceptions, such as Muller, the "gene pool" has been a distant abstraction. Eugenicists have typically emphasized immediate impacts. Whether motivated by a desire to prevent suffering or to diminish the financial burden on society—or both—the focus has been on reducing the supply of "disastrous children" in the near term.

That eugenicists have always been more concerned with mental than physical traits also helps explain the indifference to potential dysgenic effects of counseling. Much early work on the heredity of clinical diseases was pursued by eugenicists who were at least as interested in behavior as health. Charles Davenport, for example, worked simultaneously on the inheritance of Huntington's Chorea, epilepsy, a cheerful temperament, and "nomadism." The 1931 edition of the influential textbook *Human Heredity,* by Erwin Baur, Eugen Fischer, and Fritz Lenz, describes hundreds of anomalies and normal traits, some of which are today considered hereditary and some not. Like all human genetics textbooks in the 1930s and 1940s, it discussed diseases, socially aberrant behaviors, and a host of mental and temperamental characters. Its catalogue of traits included glaucoma, various cancers, Parkinson's disease, susceptibilities to rickets, hypertension, and gall stones, as well as schizophrenia, manic-depressive insanity, homosexuality, idiocy, genius, power of imagination, and talents for painting, technical invention, and science.

Most geneticists would have agreed with Lenz that "the mental differences among men are not only much greater than the physical, but also far more meaningful."[37] In Germany, the physically handicapped were subjected to sterilization and later murder. But even there, eugenicists emphasized mentality and behavior. Most sterilizations carried out under the 1933 Law for the Prevention of Progeny with Hereditary Diseases were for feeblemindedness, schizophrenia, and alcoholism. Only about one-tenth were for physical disorders.[38]

The interest in behavior carried over into the postwar period. For example, the principal projects of the Dight Institute in 1952 included mental deficiency and "normal intelligence and differential fertility" along with more obviously medical studies. Even after the American Eugenics Society began to support work in medical genetics, mentality remained its primary concern. Osborn thought that the eugenics movement should not emphasize physical health.

What really matters is a change in reproductive behavior by the intellectual elite.[39] Physical improvement will follow from any program concerned with mental qualities, which are in any case much more important. "Eugenics is particularly interested in the psychological traits of intelligence and personality, because these traits are of major importance to civilization," he explained. "If there is justification for a broad eugenics movement, it is chiefly because of the part played by heredity in providing the necessary potentials for the development of high qualities of intelligence and personality."[40]

The trait valued above all others was intelligence. In Muller's view, "for man, it is world of mental life which counts by far the most, the rest being pretty much subsidiary," while Neel claimed that "given that the most important and precious asset of any human being is his intelligence, the impact of a convincing demonstration of this on national priorities would surely exceed the conquest of a dozen rare genetic diseases."[41]

Most clinicians believed that it was the total genotype, not the single gene, that mattered. Thus, Herndon argued that "one must not only consider the obvious abnormality which prompts the patient to seek genetic advice; one must also take into account all evidence that may be obtainable concerning the total genetic potential of the parties concerned."[42] Concerning a case where the wife had surgery for harelip and cleft palate, he concluded that the couple's intellect and general genetic endowment were "sufficiently above normal that their reproduction might be advantageous to society as a whole, offsetting the disadvantage of the possible continuation of the defective gene."[43] Dice likewise believed that "the obligations of a heredity clinic will not be fully discharged . . . if it confines itself entirely to the heredity of medical pathologies." Thus, in giving advice, the geneticist should take into account "mental ability, and social worth in addition to hereditary defects."[44] According to Madge Macklin, "in dealing with these patients who ask advice, one must consider not only the fact that they have inherited diseases which they may transmit, but also whether they have highly desirable characters which they may pass on."[45]

The belief that those who availed themselves of counseling were superior in mentality and character to the general population illuminates some objections to directive counseling. Reed believed that "the counselor has never suffered the particular circumstances which the parents of the affected child suffered and therefore cannot completely understand their feelings."[46] He also assumed that individuals motivated to consult a counselor were usually well above average in character and intellect. Thus, behavior that is dysgenic with respect to a particular defective gene might still be desirable since "those people who are sufficiently concerned about their future children to come the Dight Institute for counseling have commendable concepts of their obligations as parents and these laudable characteristics should be transmitted to the next generation."[47] Harold Falls argued that the fact that they seek genetic advice indicates that the prospective parents are more intelligent and socially and morally responsible than most and reasoned that they "should actually be encouraged to have children (anticipating transmission of superior qualities) providing the gene to be transmitted does not impose too serious a handicap on the affected child."[48]

Clinicians who opposed directiveness also assumed that their clients did not need guidance to make the right choice. "From my experience in giving advice about heredity to families in all walks of life I can affirm that every parent desires his children to be free from serious handicap," wrote Dice. "If there is known to be a high probability of transmitting a serious defect, it would be an abnormal person indeed who would not refrain from having children."[49] Thus, counseling would automatically serve the interests of both individuals and society. Given adequate information, the type of middle-class people who availed themselves of genetic services would act rationally. Neel consistently condemned eugenics and opposed directive counseling—and also argued that we should use all ethical means to limit the number of those unfortunates incapable of fully participating in our complex society. In his view, genetic counseling represented one such means, "since once the principle of parental choice of a normal child is established, it seems probable that in large measure the parental desire for normal children can be relied on to result in the purely voluntary elimination of affected fetuses."[50] Cedric Carter likewise noted that most parents and patients act sensibly on the basis of the counselor's advice.[51] Or as Reed remarked, "people of normal mentality, who thoroughly understand the genetics of their problems, will behave in the way that seems correct to society as a whole."[52]

Irrational individuals were a different matter. Writing of problems in counseling individuals with phenylketonuria, Reed argued that "no couple has the right to produce a child with a 100 per cent chance of having PKU, and it is doubtful whether a couple has the right deliberately to take a 50 per cent chance of producing such a serious defect."[53] (Since the fitness of the recessive homozygote is nearly zero, there would be no effect on the "gene pool.") Thus, the case against directive counseling was based on the assumption that most families would act responsibly, not on a principle of procreative liberty.

The New Ethos of Genetic Services

Two decades later, counseling services in the United States and Canada began a rapid expansion. The first master's level program for professional counselors, at Sarah Lawrence College, was established only in 1969. At that time, about eighty percent of counselors were physicians while another eleven percent were Ph.D. geneticists.[54] While few of these counselors admitted to giving outright advice, most thought it appropriate to inform their clients in such a way as to guide them to an appropriate decision. In 1973, F. Clarke Fraser noted that, over time, he had "evolved in a more rather than less directive approach in giving genetic advice."[55] But forces were quickly building in the opposite direction.

Primary among these was the transformation in public attitudes toward reproductive responsibility that took place in the 1960s and 1970s. No longer was it assumed that society had a legitimate interest in who reproduced. Within genetic counseling, concern for the future of the population was replaced by concern for the welfare of individual families, as defined by the families themselves. That change reflected events specific to the field as well as general trends in the culture. Masters-level genetic counseling developed largely outside the field of medical genetics. Thus, none of the founders of the first program at Sarah Lawrence were geneticists.[56] In other ways as well, the new counselors were different from their predecessors. All but a handful were women, who generally value reproductive autonomy more than do men.[57] They were also trained in "client-centered" therapy, which stresses the counselor's role in clarifying the client's own feelings.

In a classic 1972 article, Claire Leonard, Gary Chase, and Barton Childs asserted that "genetic counseling is preventive medicine and should be so regarded."[58] By the end of the decade, few counselors would agree. They had rejected not just eugenics but also the public health orientation that informed the world of researchers on human metabolic disorders. These counselors rarely spoke the language of cost-benefit analysis, much less of selective breeding. They did not aim either to spare future generations' suffering or to save the state money. Instead, they hoped to empower their clients to make their own decisions according to their own values.[59] "Individual choice" and "personal autonomy" became the new catchwords. Of course, theory and practice may sometimes—or even often—diverge. But at least in North America, the shift in ethos was dramatic.

Genetic services are now everywhere justified as increasing the choices available to women. In its 1983 report, the President's Commission for the Study of Ethical Problems in Medicine and Biomedical Research identified the primary value of screening for cystic fibrosis (CF), the most common recessive genetic disease among Caucasians of European descent, as providing people with the information they would consider helpful in autonomous decision making—an aim reiterated in a 1992 report of the Office of Technology Assessment (OTA).[60] On the new view, counselors aim to serve only their clients, never society. The Professional Code of Ethics of the National Society of Genetic Counselors defines the counselor-client relationship as "based on values of care and respect for the client's autonomy, individuality, welfare, and freedom."[61] This view is likewise reflected in the 1994 recommendations of the Institute of Medicine's Committee on Assessing Genetic Risks. "The standard of care should be to support the client in making voluntary informed decisions," wrote the committee. *The goal of reducing the incidence of genetic conditions is not acceptable, since this aim is explicitly eugenics; professionals should not present any reproductive decision as 'correct' or advantageous for a person or society.*[62]

But at the level at which public policy is made, genetic services were and are still funded in hopes of reducing the incidence of genetic disease and thus saving the state money. In the 1970s, the U.S. government played a major role in promoting amniocentesis. Theodore Cooper, then assistant secretary of the Department of Health, Education, and Welfare, wrote at the time: "By focusing on prevention we increase the resources available for other programs. Few advances compare with amniocentesis in their capability for prevention of disability."[63]

Given the current bitter debate over abortion, it is unlikely that he would speak so openly today. Indeed, cross-pressures from the antiabortion movement have produced schizophrenic policies in some states. For example, Tennessee forbids the use of public funds for prenatal diagnosis of conditions for which there is no effective therapy on the grounds that abortion is against public policy— while also legislating that public funds may be used for abortion in the case of fetuses with "severe physical deformities or abnormalities, or severe mental retardation."[64] But while cost saving is often at war with other motivations, and is today rarely made explicit, it remains an important aim of genetic services programs. As the philosopher Arthur Caplan has noted, "When the state of California offers [a test] to all pregnant women it does so in the hope that some of those who are found to have children with neural tube defects will choose not to bring them to term; thereby, preventing the state from having to bear the burden of their care."[65] Economic appraisals of prenatal screening programs generally assume that benefits arise only from abortion of an affected fetus.[66]

As in the past, many presume that these individual and social interests are congruent; that families will act "rationally." Thus, policy analyses of screening programs typically presume that all identified fetuses will be aborted.[67] Today, everyone favors increasing the choices available to women. But fostering reproductive autonomy is rarely if ever the primary goal of governments when they choose to fund genetic services.[68] That states expect to save money is evident in the arguments actually made to legislatures, which are typically framed in cost-benefit terms. Thus, it seems that the new consensus on reproductive autonomy rests on the old assumption that families will ordinarily make the "right" decisions.

That assumption is questionable. As Rayna Rapp has noted, "there is no inevitable bridge between a positive diagnosis and an abortion."[69] The women she interviewed did not necessarily, or even generally, equate testing with abortion. Under hypothetical circumstances, most people are receptive to the idea of being screened. But many of those who express positive attitudes toward prenatal testing indicate that they would not abort even if the test identified a serious genetic condition.[70]

That finding is confirmed by studies of attitudes toward pregnancy termination for specific disorders. In general, they show a reluctance to abort for medical conditions except where certain early death or severe mental retardation is involved. For example, while most women are interested in knowing their CF carrier status, they are hesitant to use that information to prevent the birth of an affected child.[71] Thus, only twenty percent of parents of children with CF say they would abort an affected fetus—a higher percentage than for many other conditions, such as an incurable, severe, painful disorder that strikes at age forty.[72] Of three hundred women participating in a program of MSAFP screening for neural tube defects, seventy-one said they would refuse abortion "even if the fetus had multiple, severe handicaps such as hemiplaga and bowel and bladder incontinence."[73] Although ninety-seven percent of individuals at high risk for autosomal dominant polycystic kidney disease (a late-onset disorder that results in renal failure), and fifty percent would use prenatal testing, only eight percent would terminate a pregnancy for that reason.[74] Of course, people may behave differently when facing an actual choice than they say they would when presented with a hypothetical scenario. But the proportion of women choosing abortion is often much lower than predicted on the basis of attitude surveys.[75] The actual reluctance of women to terminate pregnancies for fetal conditions helps explain the low utilization rates for some DNA-based tests.

Many eugenicists thought that the job of ridding the world of the "unfit" could be as easily—or even better—carried out by individuals themselves. They only needed to be educated and given the tools for the job. But even women who are not opposed to abortion per se are often uneasy with *selective* abortion. Willingness to abort for fetal conditions is associated with class and ethnicity; nonwhites, and the less wealthy and educated, are more tolerant of handicaps.[76] Their resistance to selective abortion has exposed stresses that were muted when genetic counseling was a small-scale enterprise: clients were overwhelmingly white, educated, and middle-class, and prenatal diagnosis was unavailable. With the expansion and increasing diversity of the client population, it is becoming evident that reproductive choice and "public health" models of genetic services do not easily cohere.

As genetic tests become cheaper and more reliable, and increasingly applicable to common diseases (representing large markets), incentives will mount to screen more women for more disorders at an earlier age. The 1992 OTA report on the implications of CF carrier screening noted: "Without offering judgment on its appropriateness or inappropriateness, OTA finds that the matter of CF carrier screening in the United States is one of when, not if."[77] As screening programs expand, counseling is increasingly provided by obstetricians who do not fully share the professional counselors' commitment to principles of autonomous decision making and informed consent and fear becoming targets of malpractice or wrongful birth suits if they fail to test.[78] Thus, screening tests are increasingly framed as a routine part of medical care.[79] Indeed, the strongest variable in determining uptake of screening is not the attitudes of consumers but the approach taken by the health care provider; high usage is achieved by active recruitment.[80] The contradictions between autonomy and public health models is thus intensifying. How they are resolved—or suppressed—will reveal whether the contemporary consensus on reproductive autonomy is apparent or real.

Notes

1. Sheldon Reed, "A Short History of Genetic Counseling," *Dight Institute Bulletin* #14 (1974), pp. 4–5.

2. "Proposal," Rockefeller Archive Center, North Tarrytown, New York; Record Group 1.1, Series 200, Box 154, Folder 1393.

3. Sheldon Reed, "Heredity Counseling and Research," *Eugenical News* 37 (1952): 43.

4. Sheldon Reed, *Counseling in Medical Genetics* (Philadelphia: W. B. Saunders, 1955), p. 14.

5. *Ibid.*, pp. 4–15.

6. Susan Johnson, "Eugenics in the Aftermath of WWII: A Study of articles in *Science, Nature,* and the *Scientific Monthly.*" Unpublished paper, History of Eugenics seminar, University of Massachusetts at Boston, 1995.

7. Clarence P. Oliver, "A Report on the Organization and Aims of the Dight Institute," Dight Bulletin #1 (1943), p. 2.

8. Elmer Roberts, "Biology and Social Problems," *Dight Institute Bulletin* #4, (1946): 18.

9. G. Phelps, "The Eugenics Crusade of Charles Fremont Dight," *Minnesota History* 49 (1984): 99–108; Sheldon Reed, "A Short History," pp. 1–3; E. Swanson, "Biographical Sketch of Charles Fremont Dight," *Dight Institute Bulletin* #1 (1943), pp. 9–22.

10. For standard accounts of the founding of Bowman-Gray, see Ian Porter, "Evolution of Genetic Counseling in America," in H. A. Lubs and F. de la Cruz, eds., *Genetic Counseling* (New York: Raven Press, 1977), p. 26,

and Marston Meads, *The Miracle at Hawthorne Hill* (Winston-Salem: Medical Center of Bowman-Gray School of Medicine and North Carolina Baptist Hospital, 1988), p. 51, where Draper is characterized as "a New York philanthropist with a deep interest in population genetics."

11. Ruggles Gates to Sheldon Reed, August 13, 1954. Sheldon Reed papers, University of Minnesota.

12. On the Pioneer Fund, see Barry Mehler, "The New Eugenics: Academic Racism in the U.S. Today," *Science for the People* 15 (May–June, 1983): 18–23, and essays by Adam Miller, "Professors of Hate," and J. Sedgwick, "Inside the Pioneer Fund," in R. Jacoby and N. Glauberman, eds., *The Bell Curve Debate: History, Documents, Opinions* (New York: Times Books, 1994), 162–78, 144–61. On the Rockefeller Foundation, see Diane B. Paul, "The Rockefeller Foundation and the Origins of Behavior Genetics," in K. Benson et al., eds., *The Expansion of American Biology* (New Brunswick, N.J.: Rutgers University Press, 1991), pp. 262–83.

13. Ashley Montagu, *Human Heredity* (Cleveland: World Publishing, 1959), pp. 305–306.

14. Linus Pauling, "Reflections on the New Biology," *UCLA Law Review* 15 (1968): 267–72.

15. Bentley Glass, "Science: Endless Horizons or Golden Age?" *Science* 17 (1971): 28.

16. H. J. Muller, "Our Load of Mutations," *American Journal of Human Genetics* 2 (1950): 111–76; see also Diane B. Paul, "'Our Load of Mutations' Revisited," *Journal of the History of Biology* 20 (1987): 321–35.

17. Lee R. Dice, "The Structure of Heredity Counseling Services," *Eugenics Quarterly* 5 (1958): 40.

18. Lee R. Dice, "Concluding Remarks," in "A Panel Discussion: Genetic Counseling," *American Journal of Human Genetics* 4 (1952): 346.

19. Barton Childs, "Genetic Counseling: A Critical Review of the Published Literature," in B. Cohen, ed., *Genetic Issues in Public Health and Medicine* (Springfield, Ill.: Charles C. Thomas, 1978), p. 347. This is also the view of James R. Sorenson, "Genetic Counseling: Values that Have Mattered," in D. M. Bartels *et al.*, eds., *Prescribing Our Future: Ethical Challenges in Genetic Counseling* (New York: Aldine de Gruyter, 1993), pp. 3–14.

20. Clarence P. Oliver, "Human Genetics Program at the University of Texas," *Eugenical News* 37 (1952): 25–31.

21. Clarence P. Oliver, "Statement," in "A Panel Discussion: Genetic Counseling," *American Journal of Human Genetics* 4 (1953): 343.

22. C. Nash Herndon, "Human Resources from the Viewpoint of Medical Genetics," *Eugenical News* 35 (1950): 8.

23. Franz Kallmann, "Types of Advice Given by Heredity Counselors," *Eugenics Quarterly* 5 (1958): 48–50.

24. Curt Stern, *Principles of Human Genetics* (San Francisco: W. H. Freeman, 1949), p. 603.

25. I say "broadly defined" because genetics has been largely irrelevant both to the diagnosis and treatment of metabolic disorders. Thus, PKU is a genetic disease, but it is identified through a biochemical test and treated through diet.

26. Lionel S. Penrose, "Phenylketonuria: A Problem in Eugenics," *The Lancet* (June 29, 1946): 949–51.

27. Diane B. Paul and Paul J. Edelson, "The Struggle over Screening," in D. de Chadarevian and H. Kamminga, eds., *Molecularising Biology and Medicine: New Practices and Alliances, 1930s–1970s* (Reading, U.K.: Harwood Academic, 1998), pp. 203–20.

28. Richard L. Masland, "The Prevention of Mental Subnormality," in R. L. Masland *et al., Mental Subnormality: Biological, Psychological, and Cultural Factors* (New York: Basic Books, 1958), p. 15.

29. On developments in human cytogenetics, see Daniel J. Kevles, *In the Name of Eugenics: Genetics and the Uses of Human Heredity* (New York: Knopf, 1985), pp. 238–68.

30. Lawrence Snyder, "Heredity and Modern Life," in R. G. Gates *et al.*, eds., *Medical Genetics and Eugenics*, Vol. 2 (Philadelphia: Women's Medical College of Pennsylvania, 1943), p. 24.

31. William Allan *et al.*, "Some Examples of the Inheritance of Mental Deficiency: Apparently Sex-Linked Idiocy and Micro-Encephaly," *American Journal of Mental Deficiency* 48 (1944): 28.

32. For example, see Gordon Allen, "Perspectives in Population Genetics," *Eugenics Quarterly* 2 (1955): 91; James V. Neel and William Schull, *Human Heredity* (Chicago: University of Chicago Press, 1954), p. 256; Tage Kemp, "Genetic Hygiene and Genetic Counseling," *Acta Genetica et Statistica Medica* 4 (1953): 297; and Gardner Murphy, "A Research Program for Qualitative Eugenics," *Eugenics Quarterly* 1 (1954): 209–12.

33. For example, see L. L. Cavalli-Sforza and W. F. Bodmer, *The Genetics of Human Populations* (San Francisco: W. H. Freeman, 1971), pp. 757–58.

34. Frederick Osborn, Transcript, Oral History Interview (July 10, 1974), Columbia University, New York, p. 7.

35. Frederick Osborn, *The Future of Human Heredity* (New York: Harper and Brothers, 1968), p. 25.

36. E. W. MacBride, "British Eugenists and Birth Control," *Birth Control Review* 6 (1922): 247.

37. Fritz Lenz, "Die Erblichkeit der Geistigen Eigenschaften," in E. Baur *et al., Menschliche Erblichkeitslehre und Rassenhygiene*, Band I: *Menschliche Erblehre* (Munich: J. F. Lehmann, 1936), p. 661.

38. Robert Proctor, *Racial Hygiene: Medicine under the Nazis* (Cambridge, Mass.: Harvard University Press, 1988), pp. 107–108.

39. [Frederick Osborn], "Editorial," *Eugenics Quarterly* 1 (1954): 2.

40. Frederick Osborn, *Preface to Eugenics,* 2d ed. (New York: Harper and Brothers, 1951), p. 82.

41. Muller, "Our Load of Mutations," p. 165; James V. Neel, "On Emphases in Human Genetics," *Genetics* 78 (1974): 39.

42. C. Nash Herndon, "Heredity Counseling," *Eugenics Quarterly* 1 (1954): 66.

43. C. Nash Herndon, "Statement," in "A Panel Discussion: Genetic Counseling," *American Journal of Human Genetics* 4 (1952): 335.

44. Lee R. Dice, "Heredity Clinics: Their Value for Public Service and Research," *American Journal of Human Genetics* 4 (1952): 6.

45. Madge T. Macklin, "The Value of Medical Genetics to the Clinician," in C. B. Davenport *et al., Medical Genetics and Eugenics* (Philadelphia: Women's Medical College of Pennsylvania, 1940), p. 138.

46. Reed, *Counseling in Medical Genetics*, p. 339.

47. Sheldon Reed, "Heredity Counseling," *Eugenics Quarterly* 1 (1954): 48–49.

48. Harold Falls, "Consideration of the Whole Person," in H. Hammons, ed., *Heredity Counseling* (New York: Hoeber-Harper, 1959), p. 99; see also V. Cowie, "Genetic Counseling and the Changing Impact of Medical Genetics," and A. Barnes, "Prevention of Congenital Anomalies from the Point of View of the Obstetrician," in *Second International Conference on Congenital Malformations* (New York: International Medical Congress, 1964), pp. 375, 378–79.

49. Lee R. Dice, "Heredity Clinics," p. 2.

50. James V. Neel, "Lessons from a Primitive People," *Science* 170: 820–21; see also *idem, Physician to the Gene Pool* (New York: Wiley, 1994), p. 361.

51. C. O. Carter, "Prospects in Genetic Counseling," in A. Emery, ed., *Modern Trends in Human Genetics I* (New York: Appleton, 1970), pp. 340–41.

52. Sheldon Reed, "Heredity Counseling and Research," p. 43.

53. Sheldon Reed, *Parenthood and Heredity* (New York: John Wiley and Sons, 1964), p. 85.

54. J. R. Sorenson and A. J. Culbert, "Genetic Counselors and Counseling Orientation—Unexamined Topics in Evaluation," in H. A. Lubs and F. de la Cruz, eds., *Genetic Counseling* (New York: Raven Press, 1974).

55. F. Clarke Fraser, "Genetic Counseling," in V. McKusick and R. Claiborne, eds., *Medical Genetics* (New York: HP Publishing, 1973), p. 225.

56. Melissa Richter, who first suggested the program, was a physiologist and dean for the Center of Continuing Education at Sarah Lawrence; Joan Marks, the program's director, was trained as a psychiatric social worker; and Virginia Apgar, developer of the Apgar system for scoring newborns, was a teratologist. I am grateful to Robert Resta for pointing

out the disjunction between the program's founders and the major players in human genetics.

57. Dorothy Wertz and John Fletcher, "Ethical Decision Making in Medical Genetics: Women as Patients and Practitioners in Eighteen Nations," in K. Ratcliff *et al., Healing Technology: Feminist Perspectives* (Ann Arbor: University of Michigan Press, 1989), pp. 221–41.

58. C. O. Leonard *et al.,* "Genetic Counseling: A Consumers' View," *New England Journal of Medicine* 287 (1972): 437.

59. See Seymour Twiss, "The Genetic Counselor as Moral Advisor," *Birth Defects Original Articles Series 15* (1979), p. 201.

60. U.S. Congress, Office of Technology Assessment, *Cystic Fibrosis and DNA Tests: Implications of Carrier Screening,* OTA-BA-532 (Washington, D.C.: U.S. Government Printing Office, August, 1992), p. 213.

61. D. Bartels *et al.,* eds., "Code of Ethics. National Society of Genetic Counselors," in *Prescribing our Future: Ethical Challenges in Genetic Counseling* (New York: Aldine de Gruyter, 1993), p. 170.

62. Lori B. Andrews *et al.* eds., *Assessing Genetic Risks: Implications for Health and Social Policy* (Washington, D.C.: National Academy Press, 1994), pp. 14–15. Italics in original.

63. Theodore Cooper, "Implications of the Amniocentesis Registry Findings," unpublished report (October 1975), p. 2.

64. Ellen Wright Clayton, "Reproductive Genetic Testing: Regulatory and Liability Issues," in E. Thomson *et al.,* eds., *Reproductive Genetic Testing: Impact upon Women*, supplement to *Fetal Diagnosis and Therapy* 8 (Basel: Karger, 1993), pp. 54–55.

65. Arthur Caplan, "Neutrality Is Not Morality: The Ethics of Genetic Counseling," in D. Bartels *et al.,* eds., *Prescribing our Future: Ethical Challenges in Genetic Counseling* (New York: Aldine de Gruyter, 1993), p. 159.

66. Gavin Mooney and Mette Lange, "Ante-natal Screening: What Constitutes 'Benefit'?" *Social Science and Medicine* 37 (1993): 873.

67. Ruth Faden *et al.,* "Prenatal Screening and Pregnant Women's Attitude toward the Abortion of Defective Fetuses," *American Journal of Public Health* 77 (1987): 3.

68. Ruth Chadwick, "What Counts as Success in Genetic Counselling?" *Journal of Medical Ethics* 19 (1993): 43–46.

69. Rayna Rapp, "Chromosomes and Communication: The Discourse of Genetic Counseling," *Medical Anthropology Quarterly* 2 (1988): 152.

70. Eleanor Singer, "Public Attitudes toward Genetic Testing," *Population Research and Policy Review* 10 (1991): 235–55.

71. J. Botkin and S. Alemagno, "Carrier Screening for Cystic Fibrosis: A Pilot Study of the Attitudes of Pregnant Women," *American Journal of Public Health* 82 (1992): 723–25.

72. Dorothy Wertz *et al.,* "Attitudes toward Abortion among Parents of Children with Cystic Fibrosis," *American Journal of Public Health* 81 (1991): 992–96.

73. Faden *et al.,* "Prenatal Screening," p. 3.

74. Eva Sujansky *et al.,* "Attitudes of At-Risk and Affected Individuals Regarding Presymptomatic Testing for Autosomal Dominant Polycystic Kidney Disease," *American Journal of Medical Genetics* 35 (1990): 510–15.

75. For example, see A. M. Manicol *et al.,* "Implications of a Genetic Screening Programme for Polycystic Kidney Disease," *Aspects of Renal Care* 1 (1986), 219–22; M. L. Watson *et al.,* "Adult Polycystic Kidney Disease," *British Medical Journal* 300 (1990): 62–63; S. Adam *et al.,* "Five Year Study of Prenatal Testing for Huntington's Disease: Demand, Attitudes, and Psychological Assessment," *Journal of Medical Genetics* 30 (1993): 549–56; Sabine Eggers *et al.,* "Facioscapulohumeral Muscular Dystrophy: Aspects of Genetic Counseling, Acceptance of Preclinical Diagnosis, and Fitness," *Journal of Medical Genetics* 30 (1993): 589–92; Mooney and Lange, p. 875.

76. Wertz *et al.,* "Attitudes toward Abortion," pp. 992–96.

77. U.S. Congress, *Cystic Fibrosis and DNA Tests,* p. 18.

78. *Ibid.,* pp. 151–52. There are currently only about 1,000 Masters-level counselors in the United States.

79. See Nancy A. Press and Carol H. Browner, "Collective Silences, Collective Fictions: How Prenatal Testing Became Part of Routine Prenatal Care," in K. H. Rothenberg and E. J. Thomson, *Women and Prenatal Testing: Facing the Challenge of Genetic Technology* (Columbus: Ohio State University Press, 1994), pp. 201–18.

80. Hilary Bekker *et al.,* "Uptake of Cystic Fibrosis Testing in Primary Care: Supply Push or Demand Pull?" *British Medical Journal* 306 (June 1993): 1584–86.

9

GENES AND CONTAGIOUS DISEASE:
THE RISE AND FALL OF A METAPHOR

In the nineteenth century, it was assumed that mental and moral defectives were easily recognized by their physical stigmata—an assumption nicely encapsulated in the title of Daniel Pick's book, *Faces of Degeneration*.[1] This notion carried over into the eugenics movement of the twentieth century, where it was often assumed that character—both good and bad—could be read in facial characteristics. Thus, Francis Galton created photographic composites that ostensibly revealed the common traits of various human types such as criminals and aristocrats. In numerous eugenically oriented books, posters, and movies, the unfit in particular are identifiable at a glance. They are physically repulsive—with slack jaws, low brows, small heads and handle-shaped ears.[2] In 1913, an examiner at Ellis Island wrote that "facial signs" could reveal an individual's nationality, temperament, occupation, sexual habits, melancholia, feeblemindedness, and "moral obliquity," among other conditions.[3] In popular imagery, twisted minds were visually reflected in twisted bodies.

But in 1912, the premise that degenerates looked the part was challenged by the American psychologist Henry H. Goddard, who coined the term "moron" to describe the slightly retarded. Goddard stressed the concealed character of the morons' hereditary defects, a view that had frightening implications. Segregation and sterilization could diminish the threat represented by the obviously unfit. But how could degeneracy be countered if bad heredity were hidden from view?

Deborah Kallikak, the young woman whose pedigree is traced in Goddard's *The Kallikak Family,* illustrates the problem. She was attractive and charming, could sew, read, and do simple sums, play a musical instrument and breed Siamese cats. Almost no one would have guessed that she was feebleminded. Indeed, Deborah's institutional record stated: "Has no noticeable defect."[4] Pretty and eager to please but simpleminded, moron women were easily exploited by designing men. Moreover, both parents and teachers were reluctant to recognize their defects. Moron men, being physically vigorous, also easily blended in with their communities. These "high-grade" feebleminded were far more dangerous to society than were the obvious idiots and imbeciles. "While the groveling idiot is unlikely to become a parent," warned Paul Popenoe, "the moron is almost certain to do so, either legitimately or illegitimately, unless prevented by society."[5]

Goddard has been accused of deliberately doctoring photographs of the non-institutionalized members of the bad "kakos" family line in order to accentuate their sinister character.[6] But as Raymond Fancher has argued, it was probably Goddard's publisher who retouched the photographs and for purely technical reasons. Goddard's own interests would hardly have been served by making the Kallikaks appear more menacing—given that his argument for mental testing rested on the claim that morons could not be visually distinguished from the normal population.[7] Indeed, Goddard rejected Charles Davenport's request for a photograph of "an imbecile with extreme features" to be used in an exhibition on the following grounds: "I am afraid that the plan . . . for putting the face of an *evidently imbecile* child in the circle would defeat the very thing that we are struggling very hard to overcome in the popular mind, and that is the idea that these defective children show their defect in their faces. . . . I should far rather put in the circle the face of a *fine looking, normal appearing* boy or girl, and lay the emphasis on the fact that they are really feeble-minded and incapable of taking care of themselves."[8]

But *The Kallikak Family* also illustrates a more serious problem of concealment: that some cases of poor heredity may be invisible even to testers. The progenitor of the contrasting kalos and kakos lines, Martin Kallikak, Sr., was a respected member of a highly respected family. While Goddard does not make this implication explicit (perhaps, at the time, did not even see it), Martin, Sr. must have been a carrier. His dalliance with a feebleminded barmaid could not otherwise have resulted in the degenerate family line. *The Kallikak Family* warned of two dangers: from highgrade feebleminded women and from men who thoughtlessly sowed wild oats.[9] It carried the (implicit) disturbing message that mental defect might be harder to eradicate than eugenicists had thought. Thus, commenting on the Kallikak case, Popenoe noted that even in cases where feeblemindedness seems to have died out after three or four generations, "we can not be sure that the condition has not become merely latent."[10]

Goddard and Popenoe were Mendelians. But biometricians (for a very different reason) also stressed the difficulty of visually identifying the feebleminded.[11]Thus, in a 1914 paper, Karl Pearson asked "whether we can differentiate the feeble-minded from the normal by any marked physical signs," and concluded that it was impossible. "I shall believe in stigmata for the mentally defective child," he asserted, "when there has been a really scientific study of the subject; till then we may place them in the same category as the stigmata which Lambroso asserted existed in the criminal, but which Dr. Goring has so effectively demonstrated to have no existence in the case of the English convict."[12]

To Mendelians, the problem of hidden feeblemindedness became even more alarming in the late 1910s and 1920s as the implications of the Hardy-Weinberg principle were articulated. It had once been thought that preventing the affected from breeding would rapidly eradicate various undesirable traits. Geneticists now came to understand that the affected themselves represented only the tip of the iceberg. Unfortunately, most of the offending genes would be hidden in apparently asymptomatic carriers. Programs of eugenical selection would not touch this invisible reservoir, a constant source of new defectives.

Indeed, as with Martin Kallikak, the asymptomatic carriers would not themselves be aware of the risks they posed. As H. S. Jennings explained: "They are like the carriers of the typhoid bacillus that are themselves immune to the disease."[13] Like Typhoid Mary, the heterozygotes both looked and felt perfectly healthy. By the 1920s, discussions of bad heredity were often framed in the language of infectious disease; indeed, the metaphor "carrier" carries as one of its meanings the idea of unseen contagion. (It seems first to have been used in connection with heredity by Goddard in his 1914 book *Feeblemindedness*, where he writes that "this wife was at least a carrier of defect, and that defect has shown itself in Iva."[14])

Of course, infectious disease metaphors in eugenics were hardly new. Indeed, they abounded, particularly in polemics aimed at immigrants in the United States and "alien" populations in other countries. Thus, a graphic representation of the concept of hereditarily defective groups infecting the body politic appeared in the Nazi magazine *Der Sturmer;* it depicts symbols for Jews, Communists, and homosexuals seen through a microscope. The Stars of David, hammer and sickles, and triangles (as well as dollar and pound signs) appear under the heading, "Infectious Germs."[15] In the United States, immigrants were charged with carrying both genes for feeblemindedness and germs responsible for TB, typhus, cholera, and trachoma, among other epidemics.[16] And genes were often described in the language of germs, evoking the same fears of silent spread and demand for public health measures to bring it under control.

But these measures were directed at highly visible targets. It was at least in theory possible to identify and counter the threat through immigration restriction or measures to prevent despised populations from breeding. The implications of the new disease metaphors were much less obvious. How does one counter an invisible threat—especially when those who pose it, like Martin Kallikak, are white, Anglo-Saxon, and middle-class?

One possibility is for those with a particular defect in their family background to avoid mating with close relatives and with anyone belonging to a family with a history of the same defect. As early as 1910, that strategy was recommended by geneticist Charles Davenport, whose maxim was "weakness in any characteristic must be mated with strength in that characteristic; and strength may be mated with weakness."[17] Davenport's advice reflected the experience of plant and animal breeders, who knew that inbreeding was often accompanied by deterioration while the crossing of closely related strains produced an increase in growth and general "vigor."[18]

In the nineteenth century, one explanation of this phenomenon had been that parents usually possess different defects which, in crosses, tend to cancel out in the offspring. Davenport was the first geneticist to phrase this insight in Mendelian terms; in a 1908 paper, he explained that inbreeding uncovers deleterious recessives, noting that "the more foreign blood introduced the less danger of degeneration."[19]

At the time, however, Davenport did not grasp the long-term implications of this approach. Thus, in 1912 he wrote:

> The clear eugenical rule is then this: Let abnormals marry normals without trace of the defect, and let their normal offspring marry in turn into strong strains; thus the defect may never appear again. Normals from the defective strain may marry normals of normal ancestry; but must particularly avoid consanguineous marriage.

> The sociological conclusion is: Prevent the feebleminded, drunkards, paupers, sex-offenders, and criminalistic from marrying their like or cousins or any person belonging to a neuropathic strain. Practically it might be well to segregate such persons during the reproductive period for one generation. Then the crop of defectives will be reduced to practically nothing.[20]

While his proposal was subjected to ridicule by David Heron of the Galton Laboratory,[21] it is not clear that he or anyone else then understood that heterozygote carriers would vastly outnumber those actually affected, and that these carriers constituted a huge reserve from which new "defectives" would constantly arise. It was only in 1917 that E. M. East made this point and drew the implication: segregating or sterilizing the affected would not eliminate the trait. Only identifying carriers and preventing their breeding would provide a solution. Noting that complete dominance was rare among the characters studied by breeders, he speculated that mental tests could be used to identify carriers, who would likely obtain lower scores than would normal persons.[22]

It is thus not surprising that East was among the first to deem it "decidedly unwise" to follow Davenport's masking strategy.[23] By the mid-1920s, the implications of breeding by carriers was well understood. Thus, while Edwin Conklin conceded that those with slight hereditary defects might "be mated with strength in that character," he deemed it dangerous to apply the rule to serious defects. Commenting on Davenport's suggestion that a normal man could marry a feebleminded woman, he warned:

> But in all such cases the weakness is not neutralized nor removed but merely concealed in the offspring and is therefore the more dangerous. If a man chooses to marry a feebleminded woman he at least does so with his eyes open. . . . But the normal and perhaps capable children of such a union carry the taint concealed in their germplasm and if they should be mated with other normal persons carrying a similar taint some of their children would be feebleminded. . . . Such a policy of concealing weakness by mating it with strength is wholly comparable with the custom once prevalent of concealing cases of contagious diseases, and may properly be characterized as the "ostrich policy."[24]

Following a discussion of East's insight, Ruggles Gates even suggested that the tendency of carriers of feeblemindedness to intermarry is desirable since it will bring the trait "to the surface where the individual can be segregated, rather than spreading the condition subterraneously by marriage with sound stocks."[25]

In 1930, Jennings labelled the practice of masking "family eugenics." He argued that it was misguided since the defective genes would be spread and ultimately expressed; indeed, the longer the genes were hidden, the more likely was their existence to be forgotten, and individuals bearing the same recessive genes to mate.[26] Jennings considered this policy to be the greatest obstacle to "racial eugenics," which eliminates defective genes by preventing the propagation of their possessors. To really make progress, it would be necessary to identify heterozygotes. "To promote such investigations in human genetics," he wrote, "is probably now the most direct way to further the welfare of future generations through eugenic measures."[27] Many geneticists shared Jennings's hope that carriers of recessive genes could be detected through morphological, behavioral, or mental anomalies.[28]

But concrete proposals to detect heterozygotes and prevent them from breeding awaited the Nazi seizure of power. Following passage of the 1933 Law for the Prevention of Genetically Diseased Offspring, a number of German geneticists proposed practical action aimed at heterozygotes. Thus, in 1935, the half-Jewish psychiatric geneticist Franz Kallmann (who was dismissed from his position the same year) proposed sterilizing many near relatives of schizophrenics, who he thought could be identified by small anomalies.[29] In 1938, the mammalian geneticist Hans Nachtsheim noted in the text attached to a teaching film he made on genetic disease in rabbits that:

> The concept hereditarily sick is here conceived in a wider sense than in the Law for the Prevention of Genetically Diseased Offspring. Whereas according to the law only individuals who phenotypically suffer from a hereditary disease are considered hereditarily sick, here we denote all carriers of the hereditary factors of the disease as hereditarily sick. The carriers of only one recessive diseased gene, the heterozygotes, though phenotypically healthy, are not hereditarily healthy. In order to free the race from a disease, one must in an animal-breeding program eliminate the apparently healthy heterozygotes from reproduction as resolutely as is done with the sick homozygotes.[30]

But reliable techniques for identifying heterozygotes did not yet exist. By the time they did, Nachtsheim's solution was no longer fashionable.

In the 1950s, chemical methods were developed that involved either direct measurement of protein products, such as abnormal hemoglobin, or more often, measurement of enzyme activity (which is often reduced by about half in the heterozygote). It might seem that the hopes of eugenicists could, at last, be realized. But the technology was actually employed for purposes more likely to be approved by their *critics;* to reduce the immediate incidence of disease even at the price of increasing the incidence of the responsible genes. Ironically, carrier detection became a means to Jennings's "family eugenics"—the policy abandoned by Davenport when he came to understand its eugenic consequences.

That is not to say that concern with the future of the gene pool disappeared; through at least the early 1970s, some geneticists continued to worry about the dysgenic consequences of new technologies. For example, Charles Smith noted that the screening of potential parents could identify matings at risk, which could then be avoided or result in selective abortion. As a consequence, "a disease might be *temporarily* eradicated. This procedure is thus the most effective possible in the prevention of genetic disease. However, depending on the form of control used, there may be attendant undesirable changes on the population gene pool."[31]

But a more typical view was expressed by J. B. S. Haldane, who looked forward to a day when everyone would routinely undergo tests for heterozygosity at puberty so that they could avoid genetically risky matings. Haldane acknowledged that such a policy would increase the frequency of deleterious recessive genes. But he was confident that our descendants' increased knowledge of genetics would enable them to deal with the consequences.[32]

The dominant ethos of medical genetics in the 1960s is reflected in Cedric Carter's suggestion that, as biochemical methods of detecting carriers became more efficient, it would be possible to warn heterozygotes "against marrying another heterozygote, and, should two known heterozygotes marry, to screen for homozygotes *in utero,* without waiting for the birth of an affected child to indicate parents at risk";[33] Eldon Sutton's observation that "if it were possible to identify all the individuals heterozygous for a recessively inherited trait, it should be possible, by avoiding matings between two such individuals, to prevent the occurrence of diseased offspring";[34] or Merton Honeyman's and Ira Gabrielson's comment that "identification of heterozygotes and counseling of individuals to avoid consanguineous marriages or marriages with other known carriers will not reduce gene frequencies but will reduce the numbers of individuals affected with recessively determined diseases, at least for some generations. This may be an important method of control."[35] By the 1960s, few geneticists asserted a need to prevent carriers from breeding in order to root out bad genes. The new focus was on masking their effects.

There are a number of reasons why carrier screening was turned to a very different purpose than eugenicists in the inter-war period expected. First was the dawning recognition that targeting all carriers was impractical. Even in the 1920s and 1930s, when "feeble-mindedness" was the overriding concern, it was recognized that the number of carriers was very large and that "negative eugenics based upon compulsory restrictions, probably cannot be carried out on a very extensive scale."[36] But in the postwar period, when the focus shifted to disease, the scope of the problem exploded. It is surely no coincidence that J. B. S. Haldane, who did so much to publicize the implications of the Hardy-Weinberg principle, was one of the first geneticists following Davenport to propose a masking strategy. Discussing the case of PKU, then thought to produce senile dementia in heterozygotes, Haldane suggested that if this and other conditions could be detected early, it would be possible "to forbid or discourage unions of two heterozygotes,"[37] i.e., to pursue the policy that eugenicists of the 1920s and 1930s had so vehemently rejected.

In his 1946 inaugural lecture as head of the Galton Laboratory at University College, London, Lionel Penrose estimated that two of every three people were carriers of at least one serious recessive defect.[38] And he noted that even a program targeting only carriers of very severe diseases would involve huge numbers of people. Penrose also used PKU to illustrate the point. In the 1940s, the incidence of the disease was thought to be about 1 in 20,000 births in the United States and 1 in 50,000 births in Britain. To eliminate the gene in Britain, it would thus be necessary to sterilize nearly one percent of the normal population. "Only a lunatic," asserted Penrose, "would advocate such a procedure to prevent the occurrence of a handful of harmless imbeciles."[39] In fact, the incidence of PKU—and hence of heterozygote carriers—turned out to be much higher. And PKU is a relatively rare disease. Five percent of the Caucasian population carries the gene for cystic fibrosis.

If nearly everyone is a carrier, any sterilization program would necessarily be arbitrary—with the targets determined by the state of carrier detection technology. In 1949, H. J. Muller suggested that the problem was even more extreme. Muller was particularly concerned with the damage wrought by genes of *small* (rather than clinically obvious) effect; he calculated that we each carried, on average, eight of these slightly deleterious mutations. Muller himself was not discouraged by the realization that "none of us can cast stones, for we are all fellow mutants together."[40] He thought it would one day be possible to identify those individuals who carried the largest number of mutations and that they would voluntarily refrain from reproducing. But he recognized that his scheme depended on technical advances that would, in effect, allow genotypes to be "surveyed" and compared. Commenting on Muller's proposal, James Neel and William Schull noted that "eugenic procedures based on our present limited knowledge cannot help being discriminatory, in the sense that they single out for action the obviously handicapped, while failing to touch those no less handicapped but in less apparent ways."[41] Muller's own scheme avoided arbitrariness. But it required a technology that did and does not exist.

It also required a change in mores that never materialized. Muller assumed that the most genetically burdened individuals would come to feel it their duty to abstain from having children "in the interests of those [generations] who are to follow."[42] In the immediate postwar period, it was common for eugenicists to express optimism that the problem of carriers would be solved through voluntary measures, such as the "increased use of contraception by normal persons whose family history indicates that they may be carriers of serious defect."[43] But the idea of a social interest in reproductive decisions instead fell into disrepute. In the 1960s, the biochemist Linus Pauling was among the most enthusiastic proponents of heterozygote detection. Pauling proposed that every young person should have tattooed on their forehead symbols for any seriously defective recessive genes, such as those producing sickle-cell anemia and PKU. He was confident that, if this were done, carriers for the same defective gene "would recognize the situation at first sight, and would refrain from falling in love with one another." He also thought that "legislation along this line, compulsory testing for defective genes before marriage, and some form of public or semi-public display of this possession, should be adopted."[44]

Pauling explicitly acknowledged that his policy would lead to a slight increase in the future incidence of disease genes, an effort he hoped to counter, à la Muller, through an educational process aimed at convincing carriers "married to normals" to have fewer than the average number of children. But Pauling wrote at a time when it was still acceptable to urge social responsibility in reproduction. It is notable that he explicitly characterized his scheme as a form of "negative eugenics." Within a few years, to label a program "eugenics" was, *ipso facto,* to condemn it. Reproductive autonomy had become a dominant cultural value. Compare Pauling's proposal or even Haldane's 1941 comments with Michael Kaback's conclusion in a 1975 review of heterozygote screening: "The concept of mandatory genetic screening among healthy individuals for reproductive counseling has strong 'Orwellian' overtones, and . . . cannot be justified."[45]

The development of an ethos of autonomy was hardly linear. In the 1970s, it was still possible for Joseph Fletcher to warn that "there are more Typhoid Marys carrying genetic diseases than infectious diseases," and assert that no one has a right to knowingly risk passing on a genetic disease. On Fletcher's view, "testes and ovaries are communal by nature, and ethically regarded they should be rationally controlled in the social interest." Medical geneticist Joseph Dancis noted a growing feeling "among both physicians and the general public that we must be concerned not simply with ensuring the birth of a baby, but of one who will not be a liability to society, to its parents, and to itself. The 'right to be born' is becoming qualified by another right: to have a reasonable chance of a happy and useful life. This shift in attitude is shown by, among other things, the widespread movement for the reform or even the abolition of abortion laws."[46] In fact, the shift was generally in the other direction. It is a second reason why carrier detection was not used in the ways eugenicists originally envisaged. And it explains why masking strategies, such as the *Chevra Dor Yeshorim* program (where marriages within some Hasidic Jewish communities are arranged to avoid matings between carriers for Tay-Sachs disease) are rarely if ever termed "eugenics" by their proponents.[47]

Third, while Pauling and Muller took the long view, they were in a small minority. For all their talk of future generations, most eugenicists were more concerned with immediate than with distant impacts. Thus, Sheldon Reed was unconcerned about the long-term impact of genetic counseling, even though he assumed it would be dysgenic.[48] But he felt strongly that those at high risk of having a child with PKU (or any serious genetic defect) should not have children.[49] Since individuals with untreated PKU rarely do reproduce, it is clear that Reed's aim was not to protect the "gene pool," but rather to prevent disease in the here and now.

Fourth, the political popularity of eugenics depended on the fact that it was aimed at "others." Goddard would surely not have proposed segregating (much less sterilizing) the middle-class Martin Kallikak, Sr. As the emphasis shifted to disease genes, the nature of the problem changed along with its scope. Carriers of disease genes are universal. They look like you or me; indeed may *be* you or I. And that, above all, explains why the infectious disease metaphor ultimately failed to spread.

————————————————— Notes —————————————————

1. Daniel Pick, *Faces of Degeneration: A European Disorder, c. 1848-c.1918* (Cambridge: Cambridge University Press, 1989).

2. Martin S. Pernick, *The Black Stork: Eugenics and the Death of 'Defective' Babies in American Medicine and Motion Pictures* (New York and Oxford: Oxford University Press, 1996).

3. Alan M. Kraut, *Silent Travelers: Germs, Genes, and the Immigrant Menace* (New York: Basic Books, 1994), p. 62.

4. Leila Zenderland, "On Interpreting Photographs, Faces, and the Past," *American Psychologist* 43 (September 1988): 743.

5. Paul Popenoe, "Feeblemindedness," *Journal of Heredity* 6 (1915): 36.

6. Stephen Jay Gould, *The Mismeasure of Man* (New York: W. W. Norton, 1981), pp. 170–72.

7. Raymond E. Fancher, "Henry Goddard and the Kallikak Family Photographs: 'Conscious Skullduggery' or 'Whig History'?" *American Psychologist* 42 (June 1987): 588.

8. Quoted in Zenderland, "On Interpreting Photographs," p. 744.

9. Leila Zenderland, "A Sermon of New Science: *The Kallikak Family* as Eugenic Parable." Paper presented at the History of Science Society Annual Meeting, Washington, D.C., 1994.

10. Popenoe, "Feeblemindedness," p. 35.

11. To biometricians, the feebleminded simply represented one tail of a continuous distribution. I am grateful to Hamish Spencer for calling this point to my attention.

12. Karl Pearson, "Mendelism and the Problem of Mental Defect. III. On the Graduated Character of Mental Defect and on the Need for Standardizing Judgments as to the Grade of Social Inefficiency which Shall Involve Segregation," in *Questions of the Day and Fray,* no. 9 (London: Dulau, 1914), p. 20.

13. H. S. Jennings, *The Biological Basis of Human Nature* (New York: W. W. Norton, 1930), p. 234.

14. Henry H. Goddard, *Feeblemindedness: Its Causes and Consequences* (New York: Macmillan, 1914), p. 292.

15. Robert Proctor, *Racial Hygiene: Medicine under the Nazis* (Cambridge, Mass.: Harvard University Press, 1988), p. 163.

16. Kraut, *Silent Travelers.*

17. Charles B. Davenport, *Eugenics* (New York: Henry Holt, 1910), p. 25. In the same pamphlet he also wrote: "The country owes it to itself as a matter of self-preservation that every imbecile of reproductive age should be held in such restraint that reproduction is out of the question. If this proves to be impracticable then sterilization is necessary—where the life

of the state is threatened extreme measures may and must be taken" (p. 16). Davenport was not a particularly consistent thinker. But a charitable reconstruction of his view is possible on the following assumptions: imbeciles generally mate with other imbeciles, thus producing more of the same—an outcome that must be prevented. But if some are to mate, it should be with normal individuals. In these cases, dominance will solve the problem (since on his pre-Hardy-Weinberg view, the chances of two carriers mating and producing abnormal offspring are very slight). Consanguineous marriages are always to be avoided. See also Charles B. Davenport, *Heredity in Relation to Eugenics* (New York: Henry Holt, 1911), p. 257.

18. See Diane B. Paul and Barbara A. Kimmelman, "Mendel in America: Theory and Practice, 1900–1919," in R. Rainger *et al.*, eds., *The American Development of Biology* (Philadelphia: University of Pennsylvania Press, 1988), pp. 281–310, esp. pp. 298–99.

19. Charles B. Davenport, "Degeneration, Albinism, and Inbreeding," *Science* 28 (1908): 455.

20. Charles B. Davenport, "The Inheritance of Physical and Mental Traits of Man and their Application to Eugenics," in W. E. Castle *et al.*, *Heredity and Eugenics* (Chicago: University of Chicago Press, 1912), pp. 286–87.

21. David Heron, "Mendelism and the Problem of Mental Defect. I. A Criticism of Recent American Work," in *Questions of the Day and Fray*, no. 7 (London: Cambridge University Press, 1913).

22. For an extended discussion of East's argument, its extension by R. C. Punnett (who showed that the Hardy-Weinberg theorem allowed the rate of selection to be calculated quite precisely), and the controversy that followed, see Diane B. Paul and Hamish G. Spencer, "Did Eugenics Rest on an Elementary Mistake?" in this volume, pp. 117–132.

23. Edward M. East and Donald F. Jones, *Inbreeding and Outbreeding: Their Genetic and Sociological Significance* (Philadelphia: J. B. Lippincott, 1919), p. 230. (East wrote the chapter on "Man" where Davenport's maxim was criticized).

24. Edwin G. Conklin, *Heredity and Environment in the Development of Men* (Princeton: Princeton University Press, 1929), pp. 313–14.

25. R. Ruggles Gates, *Heredity and Eugenics* (London: Constable, 1923), pp. 159–60.

26. H. S. Jennings, *The Biological Basis of Human Nature* (London: Faber and Faber, 1930), p. 233.

27. *Ibid.*, p. 250.

28. For example, see R. C. Punnett, "Eliminating Feeblemindedness," *Journal of Heredity* 8 (1917): 465; Gates, *Heredity and Eugenics,* p. 159.

29. Benno Müller-Hill, *Murderous Science,* trans. George Fraser (New York: Oxford University Press, 1988), pp. 11, 28–29.

30. On Nachtsheim, see Diane B. Paul and Raphael Falk, "Scientific Responsibility and Political Context: The Case of Genetics under the Swastika," in Jane Maienschein and Michael Ruse, eds., *Biology and the Foundations of Ethics* (Cambridge: Cambridge University Press, in press).

31. Charles Smith, "Ascertaining Those at Risk in the Prevention and Treatment of Genetic Disease," in Alan E. H. Emery, ed., *Modern Trends in Human Genetics I* (New York: Appleton-Century-Crofts, 1972), p. 358; italics in original. See also G. R. Fraser and A. G. Motulsky, "Long-Term Effects of Counseling on the Gene Pool," in A. G. Motulsky, ed., *Counseling and Prognosis in Medical Genetics* (New York: Hoeber, 1969).

32. J. B. S. Haldane, "The Implications of Genetics for Human Society," in *Genetics Today: Proceedings of the XI International Congress of Genetics*, vol. 2, The Hague, September 1963. (New York: Macmillan, 1965), p. xcvi. He predicted: "In many cases the condition will prove to be curable, in a few even advantageous in new environments. And they may find out how to eliminate undesirable gametes of heterozygotes, or even to provoke back mutations to normality."

33. C. O. Carter, "Prospects in Genetic Counseling," in Alan E. H. Emery, ed., *Modern Trends in Human Genetics I* (New York: Appleton, 1970), p. 345.

34. Eldon H. Sutton, *Genes, Enzymes, and Inherited Diseases* (New York: Holt, Rinehart, and Winston, 1961), p. 93.

35. Merton Honeyman and Ira Gabrielson, "Public Health Aspects of Genetic Screening," *Human Genetics* (Birth Defects Original Article Series IV), 6 (November 1968): 103.

36. Samuel J. Holmes, *Human Genetics and Its Social Import* (New York: McGraw Hill, 1936), p. 375; see also Conklin, *Heredity and Environment*, p. 309: "It will be more difficult, perhaps an impossible thing, to apply rigidly the principles of good breeding to such persons [probable carriers] and to exclude them from reproduction."

37. J. B. S. Haldane, *New Paths in Genetics* (London: George Allen and Unwin, 1941), p. 139.

38. Lionel S. Penrose, "Phenylketonuria: A Problem in Eugenics," *The Lancet*, June 29, 1946, p. 949.

39. *Ibid.*, p. 951; see also Daniel J. Kevles, *In the Name of Eugenics: Genetics and the Uses of Human Heredity* (New York: Knopf, 1985), pp. 148–63, 176–78.

40. H. J. Muller, "Our Load of Mutations," *American Journal of Human Genetics* 2 (June 1950): 169.

41. James V. Neel and William J. Schull, *Human Heredity* (Chicago: University of Chicago Press, 1954), pp. 346–47.

42. Muller, "Our Load of Mutations," p. 151.

43. Frederick Osborn, *Preface to Eugenics* (New York: Harper and Brothers, 1951), p. 58; see also Curt Stern, *Principles of Human Genetics* (San Francisco: Freeman, 1949), p. 527.

44. Linus Pauling, "Reflections on the New Biology: Foreward," *UCLA Law Review* 15 (1968): 269.

45. Michael M. Kaback, "Heterozygote Screening for the Control of Recessive Disease," in Aubrey Milunsky, ed., *The Prevention of Genetic Disease and Mental Retardation* (Philadelphia: W. B. Saunders, 1975), p. 100.

46. Joseph Dancis, "The Prenatal Detection of Hereditary Defects," in Victor A. McKusick and Robert Claiborne, eds., *Medical Genetics* (New York: HP Publishing, 1973), p. 247.

47. Thus, medical geneticists would be very unlikely to characterize the *Chevra Dor Yeshorim* program as a "eugenic solution to a culturally perceived program" as do Lavanya Marfatia, Diana Punales-Morejon, and Rayna Rapp in "Counseling the Underserved: When an Old Reproductive Technology Becomes a New Reproductive Technology," *Birth Defects: Original Article Series* 26 (1990): 116.

48. Sheldon Reed, *Counseling in Medical Genetics* (Philadelphia: W. B. Saunders, 1955), pp. 4–15.

49. Sheldon Reed, *Parenting and Heredity* (New York: John Wiley, 1964), p. 85.

10

PKU Screening: Competing Agendas, Converging Stories

In 1963, Massachusetts became the first state to initiate mandatory genetic screening of newborns for phenylketonuria (PKU), a rare autosomal recessive disorder whose incidence in the United States, Britain, and most of Western Europe is between 1 in 11,000 and 1 in 15,000 births.[1] Although aspects of the pathogenesis and population genetics of PKU remain obscure, it has been known since the 1950s that the disease results from a defect in the enzyme phenylalanine hydroxylase, which catalyzes the conversion of phenylalanine (an essential amino acid found in most foods) to tyrosine. In the absence of therapy, phenylalanine accumulates to disastrous levels in the blood. The consequences include severe behavior problems and mental retardation. About ninety percent of those affected have IQs of less than fifty.[2]

However, the disease is treatable by a diet restricted to special phenylalanine-free foods, supplemented by a formula combining extra tyrosine with other essential amino acids and vitamins and minerals. The idea that PKU might be treatable was proposed as early as the 1930s by biochemists George Jervis and Richard Block in the United States and Lionel Penrose in Britain. In 1946, Penrose suggested in his inaugural lecture at University College, London, that its effects might be alleviated "in a manner analogous to the to the way in which a child with club-feet may be helped to walk."[3] Within a few years, studies suggested that dietary treatment would indeed bring some cognitive and behavioral improvement if treatment were initiated in early infancy and a number of screening programs were established. However, the test was not reliable until the age of six to eight weeks, after the infant had been discharged from the hospital and possibly after some degree of irreversible brain damage had occurred.

In 1961, microbiologist Robert Guthrie invented a cheap and simple blood test suitable for mass screening.[4] The conjunction of the Guthrie test with a statistical investigation validating earlier studies strengthened the existing movement to screen newborns for the disease.[5] Massachusetts quickly instituted a large-scale pilot program utilizing the Guthrie test. Within the year, and without any legislative requirement, every maternity hospital in the state was screening all newborns for PKU.[6]

But hospitals in many other states were slow to establish screening programs. In 1964, the National Association of Retarded Children (NARC), which had initially favored a legislative approach, proposed a model screening law for all states; by 1975, forty-three states had adopted mandatory programs, and ninety percent of all newborns were being tested.[7] (None mandated treatment.[8]) The laws were passed in spite of considerable opposition from the medical community and researchers in the field of human metabolism.[9] While private practitioners resented state interference with the doctor-patient relationship and feared an increase in malpractice suits, the researchers questioned the reliability of the Guthrie test and denied that enough was known about the etiology, prognosis, and management of the disease to justify mandatory programs. However, the critics provides no match for PKU clinicians and lay organizations, in particular NARC and its allies in the Children's Bureau of the Department of Health, Education, and Welfare and state health departments.[10] Today, every American state screens newborns for PKU, and usually other metabolic disorders as well.[11]

Why did such a rare condition, affecting fewer than 400 American infants a year, generate such intense activity on the part of clinicians, parents, and public officials? Paul Edelson has noted the disparity between the modest impact of PKU on the nation's health and the massive campaign pursued to control it.[12] He has also suggested a number of explanations: for scientists, PKU research made Archibald Garrod's biochemical genetics suddenly relevant to clinical medicine; for clinicians working in institutions for the retarded, it moved the study of mental deficiency into the sphere of modern scientific medicine; most important, for parents and public officials, it provided an example of a form of retardation that was treatable, indeed, that would allow affected individuals to lead normal lives. "In the face of such dramatic possibilities," he argues, "arguments regarding a lack of placebo-controlled studies, or the suggestion that PKU laws intruded on the doctor-patient relationship had little impact on legislators."[13]

Mass screening was also extended to some countries unable to provide even a minimum of medical services. Through the sale of surplus farm commodities under Public Law 480 (the Agricultural Trade Development and Assistance Act of 1954), the United States acquired foreign currencies that were made available to federal agencies for medical and scientific research and other health-related activities abroad. These funds were used to answer a question of interest to the Children's Bureau: to what extent does the incidence of PKU vary with race and ethnicity?

Because PKU is such a rare disease, some early screening programs identified few if any cases. Officials in Washington, D.C., reasoned that they had better things to do with their money and temporarily ended the program; some states threatened to follow suit. It seemed that the efficiency of screening might be increased if particular populations were targeted. But in the mid-1960s, only the general population incidence of PKU was known. In hopes of obtaining data on different racial and ethnic groups, Children's Bureau staff proposed initiating screening programs in some "PL-480" countries.[14]

They did not anticipate a positive response given the extent of unmet basic maternal and child health needs in the East European and Third World countries where most PL-480 funds were available and the obstacles to providing treatment when cases were found. Children's Bureau analysts were thus surprised to find researchers in some countries, including Poland, Yugoslavia, and Pakistan, to be enthusiastic. Researchers' eagerness to cooperate, they suggested, arose "partly from their desire to be associated with the West in something that is new and exciting, and partly from their realization that this program gives them an opportunity to develop laboratory and clinical facilities which can be used for a much broader program than the detection of one rare inborn error of metabolism."[15] Even in countries with more pressing health problems, they found, "there is a strong desire to work on special problems, which gives the workers a sense of belonging to a modern scientific community. We believe the isolation of scientists in some countries, their feeling of being passed over by the march of science, should be taken into account when we determine priorities in our cooperative programs."[16]

In the United States and many other countries, mass screening, combined with early dietary treatment, did succeed in eliminating PKU as an important cause of mental retardation. Given the dearth of successful interventions for genetic disorders, it is perhaps not surprising that PKU is so often cited "the paradigm therapeutic case" of postnatal diagnosis[17] and proof of the value of genetic medicine. It has become "not only an epitome of the application of human biochemical genetics, but also a model for so-called genetic medicine and for public health."[18] The experience with PKU is also invoked as a precedent by those who wish to expand screening programs. Thus, after noting the low incidence of PKU, *Wall Street Journal* reporters Jerry Bishop and Michael Waldholz noted that "if mass screening can be justified for . . . relatively rare genetic disorders, then screening newborns for susceptibility to such common diseases as diabetes, schizophrenia, coronary heart disease, or cancer would seem even more worthwhile."[19]

But the PKU story is more complex than these (typical) accounts suggest. Mass screening has indeed prevented retardation in tens of thousands of individuals. There is no doubt that the vast majority of affected infants are better off with treatment than they would have been in its absence. But PKU screening is not an unqualified success. Several problems were recognized in the early 1960s and explain some of the initial opposition to mandated programs (although physicians' resistance to any state dictates and their concerns about malpractice suits were also significant factors).[20] Others only emerged as mass screening became a reality. But while problems have persisted, the PKU story has become increasingly simple. What do we know of the complications? And what explains their near-invisibility outside the professional literature?

In the 1960s, opponents of mandated screening expressed concerns about the many gaps in the medical understanding of PKU, the validity of the Guthrie test, and the efficacy of the recommended dietary treatment. They argued that mandated screening was premature and likely to result in a reduced commitment to research.[21] In 1975, the Committee on Inborn Errors of Metabolism of the National Research Council admitted in a generally positive report that "screening was started, frequently under mandatory laws, when questions regarding diagnosis, prognosis, and optimal management were unanswered."[22] At the time, no one knew what proportion of infants with elevated phenylalanine levels were at risk for retardation, what level of blood phenylalanine was optimal, whether restriction of phenylalanine levels in early life would prevent retardation in infants with PKU, or whether the dietary therapy could be discontinued after brain growth was complete.[23]

The early years of newborn screening were marked by high false negative and very high false positive rates, as well as unreliable laboratory work. A 1970 survey found that for every PKU infant, nineteen who did not have the disease received an initial positive screening test—reflecting the (then unknown) fact that elevated blood phenylalanine levels may result from the relatively benign condition hyperphenylalaninemia, as well as PKU.[24] As a result, some infants who did not have the disease were treated for it, with damaging results.[25] (Today, about one in every seventy affected infants is missed, while the false positive rate is about one percent.[26])

Most of the initial problems were eventually resolved. But new ones appeared. The NRC committee noted that at the time screening became widespread, subjects had not been followed long enough to determine the extent to which therapy would prevent retardation or whether specific behavioral or cognitive problems would develop. "Yet most health professionals hailed the diet as highly effective, and there was little organized effort to determine whether, in the long, run, screening would meet its objective. Only after a few lonely but loud critical voices were raised" was an effort made to determine optimal phenylalanine levels and to measure the diet's effectiveness.[27]

These studies indicated that initial assumptions about the required length, effectiveness, ease of management, and psychosocial effects of therapy were much too sanguine. Early-treated patients with PKU, while not mentally retarded, generally have lower IQs than normally expected.[28] They do poorly in arithmetic.[29] They often experience psychological problems[30] and reduced visual perception and visual-motor skills.[31] Moreover, these patients are now advised to remain on an expensive and unappealing diet for much longer than was initially anticipated.

In the 1960s, it had been assumed that only the developing brain was vulnerable to damage.[32] But studies in the following decade revealed that IQ scores declined after removal of dietary controls.[33] Children had initially been taken off the diet as early as four years; most stopped by six.[34] Although there was (and is) no consensus on when the diet should be discontinued or relaxed, recommendations have become consistently more cautious. Researchers and clinicians now generally advise diet continuation through adolescence, and most advocate lifelong restrictions. A recent national survey of treatment centers indicates that 61 percent of programs now recommend indefinite continuation of the diet for males; 77 percent recommend this policy for females. (A decade previous, only 23 percent of programs recommended indefinite continuation for males and 42 percent for female.)[35] However, compliance is hard to obtain. The diet is boring and the formula unpalatable. Children, and especially adolescents, want to eat what their friends do. It is hard to get them to remain on the diet—and even harder to get those who have stopped to resume it.

In recent years, the problem that has received the most attention is maternal PKU. If women do not resume the diet prior to conception and maintain it throughout pregnancy, the effects on their offspring are often catastrophic, including mental retardation, microcephaly, and heart defects.[36] Before the advent of newborn screening, women with PKU bore very few children. Today, their fertility is nearly normal. Thus, screening has converted a rare occurrence into a major problem. Moreover, it is not easy to locate the at-risk adolescent girls and young women. While a few are seen regularly in PKU clinics, most discontinued the diet during childhood and have not been followed for many years.[37] About 2,700 women with PKU will be of childbearing age in the next twenty years.[38] In the absence of a remedy, all the beneficial results of screening may be neutralized by the birth of retarded children to women who have ended the diet.[39]

The issues receiving the least attention are economic. Both the formula and special foods are expensive. The cost to the *pharmacist* of a year's supply of adult formula is about $4,600. The special foods are also costly; for example, a nine-ounce can of white bread costs $3.55.[40] The flour required to make a loaf of low-protein bread runs about $6.00.[41] Less than nine ounces of spaghetti costs $3.35.[42] (There is no United States maker of low-protein pasta, which is therefore imported from Europe). These figures do not include charges for shipping and handling. Who generally pays for the dietary therapy: The states? Insurers? Individuals? What is the practical experience of those families that receive a positive screening result?

Research directed to answering these questions has been meager indeed. We know little in general about how screening "programs actually feel to those they touch."[43] We know almost nothing about how they touch people economically. In its 1975 report, the NRC Committee noted that twenty-five states provided for treatment. Of these, regulations in seven specified that it be free; in one, treatment was to be provided without charge if the doctor requested it; in ten others, if the family were "in need."[44] According to the Committee, "If all infants are to be screened, then there is an obligation to ensure that all infants discovered to have PKU receive optimal therapy. Adequate means of financing the costs of special diets and other aspects of care for families not covered by insurance and unable to pay must be a societal reponsibility."[45] But few have tried to determine how the diet is actually financed, and with what results.

There exists a huge literature on scientific and medical aspects of PKU. There are also many analyses of adolescents and their parents' disease and diet-specific knowledge, the psychosocial and cognitive effects of therapy, and the impact of the disease and diet restrictions on family functioning. A Medline search stretching back to 1966 generated almost three thousand articles primarily concerned with PKU. Only one discussed the economic impact on families.[46]

Its findings were not reassuring. A survey of three treatment centers conducted by the New York State Department of Health found that most patients who had health insurance or Medicaid coverage were unable to obtain reimbursement for the formula or special foods. Payment was denied to forty-four percent of those with health insurance policies and was covered for only ten percent of those eligible for Medicaid. A public program paid food costs for children in upstate New York but not in New York City (where only infants are covered). No financial assistance was available to adults who were ineligible for Medicaid and lacked private insurance coverage for the special goods. Many families found the costs of the special diet to be onerous. The centers' staffs "interceded for patients by appealing to private insurance carriers and to local Medicaid offices to attempt to reverse decisions which had denied reimbursement for special foods. They reported that their efforts were rarely effective."[47] (However, the inadequacy of financial support for families is typical of many chronic diseases in the United States, not a unique feature of PKU.[48])

If the PKU story is so complex, why is it often described as an unqualified success? The answer lies in the moral lessons the story is employed to teach. Both enthusiasts for genetic medicine and critics of genetic determinism have come to find the story useful. These convergent interests mean that no one has an incentive to pick up the rock and see what lies underneath.

It is obvious why advocates of genetic medicine in general, and screening programs in particular, might be inclined to a cheerful interpretation. But skeptics of testing and screening abound, particularly on the political left. They would appear to have every interest in exposing problems in the model case. They are surely sensitive to the fact that the existence of an effective therapy "does not mean that it is actually accessible to the children who need it."[49] They understand that accounts of genetic medicine tend to overestimate the benefits and obscure the costs, that on-the-ground experience often deviates from theory, and that research interests may be advanced in the guise of therapeutic programs.[50] Yet they accept self-serving and wholly abstract accounts of newborn screening. Why?

Critics of screening have their own interest in presenting PKU therapy as an unqualified success. It is a common cultural assumption that what is genetic is fixed. PKU seems to provide a dramatic example of the falsity of that assumption. Although it is an "inborn error of metabolism," a knowledge of its biochemistry enables us to limit the supply of the damaging substrate. To put is another way, PKU is a trait with a heritability of 1.0. But its expression can be drastically altered by a change in environment. PKU thus demonstrates that biology is not destiny. Joseph Alper and Marvin Natowicz note that "there is a tendency among the lay public to believe that genetic means unchangeable. This belief is false. For example, the invariably serious neurological effects of phenylketonuria . . . can be largely prevented by providing the affected newborn with a phenylalanine-restricted diet."[51]

PKU screening was transformed into a simple success story during the 1970s, when it became a weapon in the controversy over the genetics of intelligence. The efficacy of treatment provided a dramatic, decisive, and easily understood rejoinder to the argument that a high heritability of IQ rendered futile efforts to boost scholastic performance.[52] The sociobiology controversy served to reinforce this trend. PKU became the standard example of the flaws of genetic determinism. The following discussion is typical:

There is an allele that, on a common genetic background, makes a critical difference to the development of the infant in the normal environments encountered by our species. Fortunately, we can modify the environments . . . and infants can grow to full health and physical vigor if they are kept on a diet that does not contain this amino acid. So it is true that there is a "gene for PKU." Happily, it is false that the developmental pattern associated with this gene in typical environments is unalterable by changing the environment.[53]

This is true—but misleading. The reader would not suspect that the dietary regime is arduous and that adolescents and adults with PKU usually suffer some degree of behavioral and cognitive impairment.

Even critics of screening programs tend to ignore the difficulties of treatment for PKU. Thus, the authors of one otherwise skeptical (and perceptive) analysis of genetic testing criticize the "increasing preoccupation" with tests in American society.[54] They suggest in particular that testing may enhance "institutional control at the cost of individual rights."[55] PKU screening might seem an excellent illustration of their fears. Most states provide for parental objection to screening on some (usually religious) grounds. In practice, these statutes or regulations have turned out to be meaningless. Few states require that parents be told they have the right to object or even that they be informed of the test.[56] They have no effective right to refuse participation. Yet the authors use PKU only to illustrate how a disease may be "easily controlled" by changing the environment. "Though sensitivity to phenylalanine is inherited," they write, "its principal manifestation, mental retardation, depends on diet. Removing phenylalanine from the diet of afflicted individuals will avoid the serious retardation that characterizes the disease. One can, in fact, have the gene, yet with proper dietary changes never show the manifestations."[57] The moral: even accurate detection of a gene will not necessarily eliminate uncertainly about disability. Another astute critique of genetic testing includes an upbeat account of treatment for PKU in support of the (reasonable) claim that treating symptoms is preferable to correcting mutant genes.[58]

The PKU story is infinitely plastic, employed by both celebrants and skeptics of genetic medicine. But it does not serve all interests equally. In his review of *The Bell Curve,* Robert Wright notes that Richard Herrnstein and Charles Murray are wrong to conclude that "equalizing environments will have no effect" on intellectual performance, for "it turns out that if you put all infants on a diet low in the amino acid phenylalanine, the disease disappears."[59] Alas, it does not. Wright's is a generally trenchant critique of the misuse of heritability estimates. But it is time to find another illustration. Critics of genetic determinism need not counter one myth with another. There are many ways to demonstrate why "genetic" should not be equated with "fixed." But there are few genetic screening programs that lend themselves to long-term evaluation. The history of PKU screening can teach us a great deal about the social and economic realities of genetic medicine—if we let it.

Notes

1. Editorial, "Phenylketonuria Grows Up," *The Lancet* 337 (May 25, 1991): 1256.

2. Katherine L. Acuff and Ruth R. Faden, "A History of Prenatal and Newborn Screening Programs: Lessons for the Future," in Ruth Faden *et al., AIDS, Women, and the Next Generation: Towards a Morally Acceptable Public Policy for HIV Testing of Pregnant Women and Newborns* (New York: Oxford University Press, 1991), p. 64.

3. Lionel S. Penrose, "Phenylketonuria: A Problem in Eugenics," *The Lancet,* June 29, 1946, p. 951. This is a reprint of Penrose's inaugural

lecture at University College, London. Penrose actually experimented with a low-phenylalanine diet in the 1930s; see Daniel J. Kevles, *In the Name of Eugenics* (New York: Knopf, 1985), pp. 177–78. George Jervis and Richard Block also suggested the possibility of treating children with a low-phenylalanine diet in the 1930s.

4. Robert Guthrie, "Blood Screening for Phenylketonuria," *Journal of the American Medical Association* 178 (1961): 863.

5. Samuel P. Bessman and Judith P. Swazey, "Phenylketonuria: A Study of Biomedical Legislation," in Everett Mendelsohn *et al.*, eds. *Human Aspects of Biomedical Innovation* (Cambridge, Mass.: Harvard University Press, 1971), p. 53; Acuff and Faden, p. 64.

6. Acuff and Faden, p. 64.

7. Committee for the Study of Inborn Errors of Metabolism, National Research Council, *Genetic Screening: Programs, Principles, and Research* (Washington, D.C.: National Academy of Sciences, 1975), p. 23.

8. *Ibid.,* p. 50.

9. Judith P. Swazey, "Phenylketonuria: A Case Study in Biomedical Legislation," *Journal of Urban Law* 48 (1971): 883–931. See also Diane B. Paul and Paul J. Edelson, "The Struggle over Metabolic Screening," in Soraya de Chadarevian and Harmke Kamminga, eds. *Molecularising Biology and Medicine: New Practices and Alliances, 1930s–1970s* (Chur: Harwood Academic Publishers, forthcoming).

10. Bessman and Swazey, "Phenylketonuria," pp. 54–55. See also Acuff and Faden, pp. 64–65; Committee for the Study of Inborn Errors of Metabolism, pp. 44–87.

11. Lori B. Andrews, *State Laws and Regulations Governing Newborn Screening* (Chicago: American Bar Association, 1985), pp. 1–2.

12. Paul J. Edelson, "Lessons from the History of Genetic Screening in the U.S.: Policy Past, Present, and Future," in Philip Boyle and Kathleen Nolan, eds., *Setting Priorities for Genetic Services* (Washington, D.C.: Georgetown University Press, forthcoming).

13. Paul J. Edelson, "History of Genetic Screening in the United States I: The Public Debate over Phenylketonuria (PKU) Testing," abstract of paper for the American Association for the History of Medicine Meeting, New York, April 28–May 1, 1994.

14. Katherine Bain and Clara Schiffer, *Experience with the Use of PL-480 Funds in Developing PKU Programs in Foreign Countries* (Washington, D.C.: Children's Bureau, Department of Health, Education, and Welfare, 1966).

15. *Ibid.,* p. 7.

16. *Ibid.,* p. 13.

17. Daniel J. Kevles, *In the Name of Eugenics* (New York: Knopf, 1985), p. 254.

18. Charles R. Scriver, "Phenylketonuria—Genotypes and Phenotypes," *New England Journal of Medicine,* May 2, 1991, p. 1280. See also Marvin R. Natowicz and Joseph S. Alper, "Genetic Screening: Triumphs, Problems, and Controversies," *Journal of Public Health Policy* 12 (1991): 479; Collen G. Azen *et al.,* "Intellectual Development in 12-Year-Old Children Treated for Phenylketonuria," *American Journal of Diseases of Children* 145 (January 1991): 35.

19. Jerry E. Bishop and Michael Waldholz, *Genome* (New York: Simon and Schuster, 1990), pp. 18–19.

20. Committee for the Study of Inborn Errors of Metabolism, National Research Council, *Genetic Screening,* pp. 46, 50.

21. Samuel P. Bessman, "Legislation and Advances in Medical Knowledge—Acceleration or Inhibition?" *Journal of Pediatrics* 69: 337.

22. Committee for the Study of Inborn Errors of Metabolism, National Research Council, *Genetic Screening,* pp. 32–40. See also Mark Lappe, *Genetic Politics* (New York: Simon and Schuster, 1979), pp. 92–93.

23. Committee for the Study of Inborn Errors of Metabolism, pp. 28–29.

24. *Ibid.,* p. 34.

25. Neil A. Holtzman, *Proceed with Caution: Predicting Genetic Risks in the Recombinant DNA Era* (Baltimore: Johns Hopkins University Press, 1989), p. 5.

26. C. C. Mabry, "Phenylketonuria: Contemporary Screening and Diagnosis," *Annals of Clinical and Laboratory Science* 6 (November–December 1990): 393–97.

27. Committee for the Study of Inborn Errors of Metabolism, National Research Council, *Genetic Screening,* pp. 88–89.

28. Harvey L. Levy, "Nutritional Therapy in Inborn Errors of Metabolism," in Robert J. Desnick, *Treatment of Genetic Diseases* (New York: Churchill Livingstone, 1991), p. 16.

29. Azen *et al.,* "Intellectual Development," pp. 38–39.

30. J. Weglage *et al.,* "Psychological and Social Findings in Adolescents with Phenylketonuria," *European Journal of Pediatrics* 151 (July 1992): 522–25.

31. K. Fishler *et al.,* "School Achievement in Treated PKU Children," *Journal of Mental Deficiency Research* 33 (December 1989): 493–98.

32. Editorial, "Phenylketonuria Grows Up," p. 1256.

33. B. Cabalska *et al.,* "Termination of Dietary Treatment in Phenylketonuria," *European Journal of Pediatrics* 126 (1977): 253–62; I. Smith *et al.,* "Effect of Stopping Low-Phenylalanine Diet on Intellectual Progress of Children with Phenylketonuria," *British Medical Journal* 2 (1978): 723–26. See also M. G. Beasley, "Effect on Intelligence of Relaxing the Low Phenylalanine Diet in Phenylketonuria," *Archives of Disease in Childhood* 66 (March 1991): 311–16.

34. Virginia E. Schuett *et al.,* "Diet Discontinuation Policies and Practices of PKU Clinics in the United States," *American Journal of Public Health* 70 (1980): 498.

35. Virginia E. Schuett, *National Survey of Treatment Programs for PKU and Selected Other Inherited Metabolic Disorders* (Rockville, Md.: Bureau of Maternal and Child Health and Resources Development. Public Health Services. U. S. Dept. of Health and Human Services, 1990), p. 11.

36. Susan Waisbren *et al.,* "The New England Maternal PKU Project: Identification of At-Risk Women," *American Journal of Public Health* 78 (July 1988): 789–92.

37. *Ibid.,* p. 789.

38. Azen, "Intellectual Development," p. 39.

39. The potential "rebound effect" was first calculated by N. H. Kirkman, "Projections of a Rebound in Frequency of Mental Retardation from Phenylketonuria," *Applied Research in Mental Retardation* 3 (1982): 319–28.

40. "Price List and Order Form," Dietary Specialties, Inc.

41. Betsy A. Lehman, "State Will Aid Those with Enzyme Deficiency," *Boston Globe*, January 12, 1994, p. 16.

42. "Price List and Order Form."

43. Ellen Wright Clayton, "Screening and Treatment of Newborns," *Houston Law Review* 29 (Spring 1992): 94.

44. Committee for the Study of Inborn Errors of Metabolism, National Research Council, *Genetic Screening,* p. 54.

45. *Ibid.,* p. 92.

46. Bernard N. Millner, "Insurance Coverage of Special Foods Needed in the Treatment of Phenylketonuria," *Public Health Reports* 108 (January–February 1993): 60–65.

47. *Ibid.,* p. 64.

48. This point was suggested by Norman Fost, personal communication, February 23, 1996.

49. Clayton, "Screening and Treatment of Newborns," p. 101.

50. See Lappe, *Genetic Politics,* p. 92.

51. Joseph S. Alper and Marvin R. Natowicz, "On Establishing the Genetic Basis of Mental Disease," *Trends in Neurosciences* 16 (1993): 387–89.

52. For example, see essays by Carl Bereiter, "Genetics and Educability: Educational Implications of the Jensen Debate," and by N. J. Block and Gerald Dworkin, "IQ, Heritability, and Inequality" in N. J. Block and Gerald Dworkin, *The IQ Controversy* (New York: Pantheon, 1976), pp. 395–96, 489; Steven Rose, "Environmental Effects on Brain and Behaviour," in Ken Richardson *et al., Race, Culture and Intelligence* (Baltimore: Penguin Books, 1972), p. 135.

53. Philip Kitcher, *Vaulting Ambition: Sociobiology and the Quest for Human Nature* (Cambridge, Mass.: MIT Press, 1985), p. 128.

54. Dorothy Nelkin and Laurence Tancredi, *Dangerous Diagnostics: The Social Power of Biological Information* (New York: Basic Books, 1989), p. 160.

55. *Ibid.,* p. 161.

56. Lori B. Andrews, *State Laws and Regulations Governing Newborn Screening* (Chicago: American Bar Association, 1985), p. 2.

57. Nelson and Tancredi, pp. 41–42.

58. Ruth Hubbard and Elijah Wald, *Exploding the Gene Myth* (Boston: Beacon Press, 1993), pp. 198–99.

59. Robert Wright, "Dumb Bell," *New Republic,* January 2, 1995, p. 6.

REFERENCES

Abir-Am, Pnina. "The Disclosure of Physical Power and Biological Knowledge in the 1930s: A Reappraisal of the Rockefeller Foundation's 'Policy' in Molecular Biology." *Social Studies of Science* 12 (1982): 341–82.

Acuff, Katherine L., and Ruth R. Faden. "A History of Prenatal and Newborn Screening Programs: Lessons for the Future." In Ruth Faden *et al. AIDS, Women, and the Next Generation: Towards a Morally Acceptable Public Policy for HIV Testing of Pregnant Women and Newborns.* New York: Oxford University Press, 1991.

Adam, S., *et al.* "Five Year Study of Prenatal Testing for Huntington's Disease: Demand, Attitudes, and Psychological Assessment." *Journal of Medical Genetics* 30 (1993): 549–56.

Adams, Mark. *The Wellborn Science: Eugenics in Germany, France, Brazil, and Russia.* New York: Oxford University Press, 1990.

Allan, William, *et al.* "Some Examples of the Inheritance of Mental Deficiency: Apparently Sex-Linked Idiocy and Microencephaly." *American Journal of Mental Deficiency* 48 (1944): 325–34.

Allen, Garland E. "The Eugenics Record Office at Cold Spring Harbor, 1910–1940: An Essay in Institutional History." *Osiris* 2 (1986): 225–64.

———. *Thomas Hunt Morgan: The Man and His Science.* Princeton: Princeton University Press, 1978.

Allen, Gordon. "Perspectives in Population Genetics." *Eugenics Quarterly* 2 (1955): 90–97.

Alper, Joseph S., and Marvin R. Natowicz. "On Establishing the Genetic Basis of Mental Disease." *Trends in Neurosciences* 16 (1993): 387–89.

Althusser, Louis. *For Marx.* London: Allen Lane, 1969.

187

Andrews, Lori B. *State Laws and Regulations Governing Newborn Screening.* Chicago: American Bar Association, 1985.

Andrews, Lori B., *et al.,* eds. *Assessing Genetic Risks: Implications for Health and Social Policy.* Washington, D.C.: National Academy Press, 1994.

Azen, Collen G., *et al.* "Intellectual Development in 12-Year-Old Children Treated for Phenylketonuria." *American Journal of Diseases of Children* 145 (January 1991): 35–39.

Aveling, Edward. *Darwinism and Small Families.* London: printed by Annie Besant and Charles Bradlaugh, 1882.

———. *Progress* 2 (1883): 210–17.

Baber, Ray Erwin. [Review of H. J. Muller, *Out of the Night.*] *American Sociological Review* 1 (1936): 533.

Bain, Katherine, and Clara Schiffer. *Experience with the Use of PL-480 Funds in Developing PKU Programs in Foreign Countries.* Washington, D.C.: Children's Bureau, Department of Health, Education, and Welfare, 1966.

Bannister, Robert. *Social Darwinism: Science and Myth in Anglo-American Social Thought.* Philadelphia: Temple University Press, 1979.

Barker, D. "The Biology of Stupidity: Genetics, Eugenics, and Mental Deficiency in the Inter-War Years." *British Journal of the History of Science* 22 (1989): 347–75.

Barnes, A. "Prevention of Congenital Anomalies from the Point of View of the Obstetrician." In *Second International Conference on Congenital Malformations.* New York: International Medical Congress, 1964. Pp. 378–79.

Bartels, D., *et al.,* eds. "Code of Ethics. National Society of Genetic Counselors." In *Prescribing our Future: Ethical Challenges in Genetic Counseling.* New York: Aldine de Gruyter, 1993.

Bateson, William. "Presidential Address to the British Association, Australia" (Sydney, Australia, [1914]). Reprinted in *William Bateson, F.R.S.: His Essays and Addresses.* New York: Garland, 1984. Pp. 297–316.

Bayer, Ronald. *Private Acts, Social Consequences: AIDS and the Politics of Public Health.* New Brunswick, N.J.: Rutgers University Press, 1989.

Beasley, M. G. "Effect on Intelligence of Relaxing the Low Phenylalanine Diet in Phenylketonuria." *Archives of Disease in Childhood* 66 (March 1991): 311–16.

Beatty, John. "Weighing the Risks: Stalemate in the 'Classical/Balance' Controversy." *Journal of the History of Biology* 20 (1987): 289–319.

Bekker, Hilary, *et al.* "Uptake of Cystic Fibrosis Testing in Primary Care: Supply Push or Demand Pull?" *British Medical Journal* 306 (June 1993): 1584–86.

Bellamy, Edward. *Looking Backward.* New York: Signet, 1888; rpt. 1960.

Bennett, J. H. *Introduction to Natural Selection, Heredity, and Eugenics.* Oxford: Clarendon Press, 1983.

Bereiter, Carl. "Genetics and Educability: Educational Implications of the Jensen Debate." In N. J. Block and Gerald Dworkin. *The IQ Controversy.* New York: Pantheon, 1976. Pp. 395–96.

Berlin, Isaiah. "Two Concepts of Liberty." In *Four Essays on Liberty.* New York: Oxford University Press, 1969. Pp. 118–172.

Bernal, J. D. *The World, The Flesh, and the Devil.* Bloomington: Indiana University Press, 1969; orig. ed. 1928.

Bessman, Samuel P. "Legislation and Advances in Medical Knowledge— Acceleration or Inhibition?" *Journal of Pediatrics* 69: 334–38.

Bessman, Samuel P., and Judith P. Swazey. "Phenylketonuria: A Study of Biomedical Legislation." In Everett Mendelsohn *et al.,* eds. *Human Aspects of Biomedical Innovation.* Cambridge, Mass.: Harvard University Press, 1971. Pp. 49–76.

Bishop, Jerry E., and Michael Waldholz. *Genome.* New York: Simon and Schuster, 1990.

Bix, Amy Sue. "Pavlovian Science Comes to America: Experimental Research of W. Horsley Gantt at the Johns Hopkins University." Unpublished manuscript, 1988.

Block, Ned. "Race, Genes, and IQ." *Boston Review of Books,* December 1995, pp. 30–35.

Block, N. J., and Gerald Dworkin. "IQ, Heritability, and Inequality." In N. J. Block and Gerald Dworkin. *The IQ Controversy.* New York: Pantheon, 1976. P. 489.

Bloor, David. *Knowledge and Social Imagery.* London: Routledge and Kegan Paul, 1976.

Blum, Debra E. "Authors, Publishers Seek to Raise Quality and Status of College Textbooks, Long an Academic Stepchild." *Chronicle of Higher Education,* July 31, 1991, pp. A11–12.

Bodmer, W. E., and L. L. Cavalli-Sforza. "Intelligence and Race." *Scientific American* 223 (1970): 19–29.

Botkin, J., and S. Alemagno. "Carrier Screening for Cystic Fibrosis: A Pilot Study of the Attitudes of Pregnant Women." *American Journal of Public Health* 82 (1992): 723–25.

Brewer, Herbert. "Eutelegenesis." *Eugenics Review* 27 (1935): 121–26.

Broberg, G., and N. Roll-Hansen, eds. *Eugenics and the Welfare State: Sterilization Policy in Denmark, Sweden, Norway, and Finland.* East Lansing: Michigan State University Press, 1996.

Broberg, G., and M. Tydén. "Eugenics in Sweden: Efficient Care." In G. Broberg and N. Roll-Hansen, eds. *Eugenics and the Welfare State: Sterilization Policy in Denmark, Sweden, Norway, and Finland.* East Lansing, Michigan State University Press, 1996. Pp. 77–149.

Brown, Theodore M. "Alan Gregg and the Rockefeller Foundation's Support of Franz Alexander's Psychosomatic Research." *Bulletin of the History of Medicine* 61 (1987): 155–82.

Caplan, Arthur. "Neutrality Is Not Morality: The Ethics of Genetic Counseling." In D. Bartels *et al.*, eds. *Prescribing our Future: Ethical Challenges in Genetic Counseling.* New York: Aldine de Gruyter, 1993. Pp. 149–65.

Carlson, A. J. [Review of H. J. Muller, *Out of the Night.*] *American Journal of Sociology* 42 (1936): 134.

Carlson, Elof Axel. *Genes, Radiation, and Society: The Life and Work of H. J. Muller.* Ithaca: Cornell University Press, 1981.

————. *Human Genetics.* Lexington, Mass.: D. C. Heath and Company, 1984.

————. "Ramifications of Genetics." *Science,* April 25, 1986, pp. 531–32.

Carter, C. O. "Prospects in Genetic Counseling." In Alan E. H. Emery, ed. *Modern Trends in Human Genetics I.* New York: Appleton, 1970. Pp. 339–49.

Castle, William E. *Genetics and Eugenics.* Cambridge, Mass.: Harvard University Press, 1927.

Castle, William E., *et al. Heredity and Eugenics.* Chicago: University of Chicago Press, 1912.

Cavalli-Sforza, L. L., and W. F. Bodmer. *The Genetics of Human Populations.* San Francisco: W. H. Freeman, 1971.

Central Association for Mental Welfare. "Sterilisation and Mental Deficiency." *Studies in Mental Inefficiency* 4 (July 15, 1923): 12.

Cabalska, B., *et al.* "Termination of Dietary Treatment in Phenylketonuria." *European Journal of Pediatrics* 126 (1977): 253–62.

Chadwick, Ruth. "What Counts as Success in Genetic Counselling?" *Journal of Medical Ethics* 19 (1993): 43–46.

Chesterton, G. K. "To the Reader." In *Eugenics and Other Evils.* London: Cassell, 1922.

Childs, Barton. "Genetic Counseling: A Critical Review of the Published Literature." In B. Cohen, ed. *Genetic Issues in Public Health and Medicine.* Springfield, Ill.: Charles C. Thomas, 1978.

Clapperton, Jane. *Scientific Meliorism.* London: K. Paul, Trench, 1885.

Clarke, Angus. "Genetics, Ethics, and Audit." *Lancet* 335 (May 12, 1990): 1145–47.

Clayton, Ellen Wright. "Reproductive Genetic Testing: Regulatory and Liability Issues." In E. Thomson *et al.*, eds. *Reproductive Genetic Testing: Impact upon Women.* Supplement to *Fetal Diagnosis and Therapy* 8. Basel: Karger, 1993.

————. "Screening and Treatment of Newborns." *Houston Law Review* 29 (Spring 1992): 85–148.

Collins, Harry M. "An Empirical Relativist Programme in the Sociology of Scientific Knowledge." In K. Knorr-Cetina and M. Mulkay, eds. *Science Observed: Perspectives on the Social Study of Science.* London: Sage, 1983.

Collins, Harry M., and Trevor Pinch. *The Golem: What Everyone Should Know about Science.* Cambridge: Cambridge University Press, 1993.

Committee for the Study of Inborn Errors of Metabolism, National Research Council. *Genetic Screening: Programs, Principles, and Research.* Washington, D.C.: National Academy of Sciences, 1975.

Conklin, Edwin G. *Heredity and Environment in the Development of Man.* 2d ed. Princeton: Princeton University Press, 1916; rpt. 1929.

————. "The Purposive Improvement of the Human Race." In *Human Biology and Population Improvement.* Ed. E. V. Cowdry. New York: Hoeber, 1930. Pp. 566–88.

Cooper, Theodore. "Implications of the Amniocentesis Registry Findings." Unpublished report, October 1975.

Cornwell, David, and Sandy Hobbs. "The Strange Saga of Little Albert." *New Society,* March 18, 1976, pp. 602–604.

Coser, Lewis A., Charles Kadushin, and Walter W. Powell. *Books: The Culture and Commerce of Publishing.* New York: Basic Books, 1982.

Cowan, Ruth Schwartz. "Genetic Technology and Reproductive Choice: An Ethics for Autonomy." In Daniel J. Kevles and Leroy Hood, eds. *The Code of Codes: Scientific and Social Issues in the Human Genome Project.* Cambridge, Mass.: Harvard University Press, 1992. Pp. 318–52.

Cowie, V. "Genetic Counseling and the Changing Impact of Medical Genetics." In *Second International Conference on Congential Malformations.* New York: International Medical Congress, 1964. P. 375.

Cynkar, J. *"Buck v. Bell:* 'Felt Necessities' v. 'Fundamental Values'?" *Columbia Law Review* 81 (1981): 1418–61.

Dancis, Joseph. "The Prenatal Detection of Hereditary Defects." In Victor A. McKusick and Robert Claiborne, eds. *Medical Genetics.* New York: HP Publishing, 1973.

Davenport, Charles B. *Eugenics.* New York: Henry Holt, 1910.

————. "Degeneration, Albinism, and Inbreeding." *Science* 28 (1908): 445–55.

————. *Heredity in Relation to Eugenics.* New York: Henry Holt and Co., 1911.

————. "The Inheritance of Physical and Mental Traits of Man and Their Application to Eugenics." In *Heredity and Eugenics,* ed. W. E. Castle, J. M. Coulter, C. B. Davenport, E. M. East, and W. L. Porter. Chicago: University of Chicago Press, 1912. Pp. 269–88.

Davis, Joel. *Mapping the Code: The Human Genome Project and the Choices of Modern Science.* New York: Wiley, 1991.

DeParle, Jason. "Daring Research or 'Social Science Pornography'?" *New York Times,* October 9, 1994, pp. 48–54, 70–80.

Dice, Lee R. "Concluding Remarks." In "A Panel Discussion: Genetic Counseling." *American Journal of Human Genetics* 4 (1952): 345–46.

————. "Heredity Clinics: Their Value for Public Service and Research." *American Journal of Human Genetics* 4 (1952): 1–13.

————. "The Structure of Heredity Counseling Services." *Eugenics Quarterly* 5 (1958): 40.

Dobzhansky, Theodosius. *Evolution, Genetics, and Man.* New York: John Wiley and Sons, 1955.

————. "Is Genetic Diversity Compatible with Human Equality?" *Social Biology* 20 (1973): 280–88.

Dunn, John. *Western Political Theory in the Face of the Future.* Cambridge: Cambridge University Press, 1979.

Duster, Troy. *Backdoor to Eugenics.* New York: Routledge, 1990.

East, Edward M. "Hidden Feeblemindedness." *Journal of Heredity* 8 (1917): 215–17.

East, Edward M., and Donald F. Jones. *Inbreeding and Outbreeding: Their Genetic and Sociological Significance.* Philadelphia: J. B. Lippincott, 1919.

Edelson, Paul J. "History of Genetic Screening in the United States I: The Public Debate over Phenylketonuria (PKU) Testing." Abstract of paper for the American Association for the History of Medicine Meeting, New York, April 28–May 1, 1994.

————. "Lessons from the History of Genetic Screening in the U.S.: Policy Past, Present, and Future." In Philip Boyle and Kathleen Nolan, eds. *Setting Priorities for Genetic Services.* Washington, D.C.: Georgetown University Press, forthcoming.

Editorial. "Phenylketonuria Grows Up." *The Lancet* 337 (May 25, 1991): 1256–57.

Eggers, Sabine, *et al.* "Facioscapulohumeral Muscular Dystrophy: Aspects of Genetic Counseling, Acceptance of Preclinical Diagnosis, and Fitness." *Journal of Medical Genetics* 30 (1993): 589–92.

Ellis, Havelock. *The Problem of Race Regeneration.* London: Cassell, 1911.

————. "The Sterilization of the Unfit." *Eugenics Review,* October 1909, pp. 203–206.

————. *The Task of Social Hygiene.* London: Constable, 1912.

Erlenmyer-Kimling, L., and L. F. Jarvik. "Genetics and Intelligence: A Review." *Science* 142 (1963): 1477–79.

"Ethical Issues Policy Statement on Huntington's Disease Molecular Genetics Predictive Test." *Journal of Medical Genetics* 27 (1990): 34–38.

Faden, Ruth, *et al.* "Prenatal Screening and Pregnant Women's Attitude toward the Abortion of Defective Fetuses." *American Journal of Public Health* 77 (1987): 288–90.

Falconer, D. S. *Introduction to Quantitative Genetics.* 3d ed. Harlow, G.B.: Longman, 1989.

Falls, Harold. "Consideration of the Whole Person." In H. Hammons, ed. *Heredity Counseling.* New York: Hoeber-Harper, 1959.

Fancher, Raymond E. "Henry Goddard and the Kallikak Family Photographs: 'Conscious Skullduggery' or 'Whig History'?" *American Psychologist* 42 (June 1987): 585–90.

Farrall, Lindsay. "The History of Eugenics: A Bibliographical Review." *Annals of Science* 36 (1979).

Fine, Beth A. "The Evolution of Non-Directiveness in Genetic Counseling and Implications for the Human Genome Project." In Dianne M. Bartels *et al.*, eds. *Prescribing Our Future: Challenges in Genetic Counseling.* Hawthorne, N.Y.: Aldine de Gruyter, 1993. Pp. 101–17.

Fish, Stanley. "Professor Sokal's Bad Joke." *New York Times,* May 21, 1996, p. A23.

Fisher, R. A. "The Elimination of Mental Defect." *Eugenics Review* 26 (1924): 114–16. Reprinted in *Collected Papers of R. A. Fisher.* Vol. 1. Ed. J. H. Bennett. Adelaide, Aus.: University of Adelaide, 1971.

Fishler, K., *et al.* "School Achievement in Treated PKU Children." *Journal of Mental Deficiency Research* 33 (December 1989): 493–98.

Flax, Jane. "The End of Innocence." In J. Butler and J. W. Scott, eds. *Feminists Theorize the Political.* London: Routledge, 1992.

Fletcher, Joseph. "Knowledge, Risk, and the Right to Reproduce: A Limiting Principle." In Aubrey Milunsky and George J. Annas, eds. *Genetics and the Law II.* New York: Plenum Press, 1980.

Fletcher, Ronald. *Science, Ideology, and the Media: The Burt Scandal.* New Brunswick, N.J.: Transaction Publishers, 1991.

Fosdick, Raymond B. *The Story of the Rockefeller Foundation.* New York: Harper & Brothers, 1952.

Fox-Genovese, Elizabeth. *Feminism without Illusions.* Chapel Hill: University of North Carolina Press, 1991.

Fraser, F. Clarke. "Genetic Counseling." In V. McKusick and R. Claiborne, eds. *Medical Genetics.* New York: HP Publishing, 1973.

Fraser, G. R., and A. G. Motulsky. "Long-Term Effects of Counseling on the Gene Pool." In A. G. Motulsky, ed. *Counseling and Prognosis in Medical Genetics.* New York: Hoeber, 1969.

Freeden, Michael. "Eugenics and Progressive Thought: A Study in Ideological Affinity." *The Historical Journal* 22 (1979): 645–71.

Fuller, John L., and William Thompson. *Behavior Genetics.* New York: Wiley, 1960.

Galton, Francis. *English Men of Science: Their Nature and Nurture.* London: Macmillan, 1874.

———. *Hereditary Genius.* London: Macmillan, 1869.

———. "Hereditary Talent and Character." *Macmillan's Magazine* 12 (1865): 157–66, 318–27.

———. *Inquiries into the Human Faculty and Its Development.* New York: Macmillan, 1883.

Gates, R. Ruggles. *Heredity and Eugenics.* London: Constable, 1923.

Geiger, Roger L. *To Advance Knowledge: The Growth of American Research Universities, 1900–1940*. New York: Oxford University Press, 1986.

"General Discussion: Mental Deficiency in Its Social Aspects." *British Medical Journal*, April 11, 1923, p. 231.

Gill, Rosalind. "Power, Social Transformation, and the New Determinism: A Comment on Grint and Woolgar." *Science, Technology, and Human Values* 21 (1996): 347–53.

Gillie, Oliver. *Sunday Times* (London), October 24, 1979, p. 1.

Glass, Bentley. "Science: Endless Horizons or Golden Age?" *Science* 17 (1971): 28.

Goddard, H. H. *Feeblemindedness: Its Causes and Consequences*. New York: Macmillan, 1914.

———. *The Kallikak Family: A Study in the Heredity of Feeblemindedness*. New York: Macmillan, 1912.

———. "Ungraded Classes." Report on Educational Aspects of the Public School Systems of the City of New York, Part II, Subdivision I, Section E. City of New York, 1911–12.

Gordon, Linda. *Woman's Body, Woman's Right: A Social History of Birth Control in America*. New York, 1976.

Gornick, Thomas W. *Book Publishing Annual (1984)*. New York: R. R. Bowker, 1984.

Gould, Stephen Jay. *The Mismeasure of Man*. New York: W. W. Norton, 1981.

Graham, Loren. "Science and Values: The Eugenics Movement in Germany and Russia in the 1920s." *American Historical Review* 82 (1977): 1133–64.

Grant, Madison. *The Passing of the Great Race*. New York: Charles Scribner's, 1916.

Green, Martin. "The Visible College in British Science." *The American Scholar* 47 (1977/78): 105–17.

Gross, Peter, and Norman Levitt. *Higher Superstition: The Academic Left and Its Quarrels with Science*. Baltimore: Johns Hopkins University Press, 1994.

Guthrie, Robert. "Blood Screening for Phenylketonuria." *Journal of the American Medical Association* 178 (1961): 863.

Haldane, J. B. S. "Biology and Marxism." *Modern Quarterly* 3 (1948): 9.

———. *Daedalus: On Science and the Future*. London: K. Paul, Trench, Trubner, 1924.

———. "Darwin on Slavery." *Daily Worker* (London), November 14, 1949.

———. *Heredity and Politics*. New York: W. W. Norton, 1938.

———. "The Implications of Genetics for Human Society." In *Genetics Today: Proceedings of the XI International Congress of Genetics. The Hague, September, 1963*. Vol. 2. New York: Macmillan, 1965. Pp. xci–cii.

———. *The Inequality of Man and Other Essays.* London: Chatto and Windus, 1928, rpt. 1932.

———. *New Paths in Genetics.* London: George Allen and Unwin, 1941.

———. "The Proper Application of the Knowledge of Human Genetics." In M. Goldsmith and A. Mackay, eds. *The Science of Science.* London: Souvenir Press, 1964. Pp. 150–56.

Haller, Mark H. *Eugenics: Hereditarian Attitudes in American Thought.* New Brunswick, N.J.: Rutgers University Press, 1984.

Harris, Ben. "Whatever Happened to Little Albert?" *American Psychologist* 34 (1979): 151–60.

Harwood, Jonathan. "Nature, Nurture and Politics: A Critique of the Conventional Wisdom." In J. V. Smith and D. Hamilton, eds. *The Meritocratic Intellect.* Aberdeen: Aberdeen University Press, 1980. Pp. 115–28.

Herndon, C. Nash. "Heredity Counseling." *Eugenics Quarterly* 1 (1955): 64–66.

———. "Human Resources from the Viewpoint of Medical Genetics." *Eugenical News* 35 (1950): 8.

———. "Statement." In "A Panel Discussion: Genetic Counseling." *American Journal of Human Genetics* 4 (1952): 335.

Heron, D. "Mendelism and the Problem of Mental Defect. I. A Criticism of Recent American Work." In *Questions of the Day and of the Fray. No. 7.* London: Cambridge University Press, 1913.

Herrnstein, Richard. "I.Q." *Atlantic* 228 (1971): 63–64.

———. *I.Q. in the Meritocracy.* Boston: Little, Brown, 1973.

Herrnstein, Richard J., and Charles Murray. *The Bell Curve: Intelligence and Class Structure in American Life.* New York: Free Press, 1994.

Hilgard, E. R., and D. G. Marquis. *Conditioning and Learning.* New York: Appleton-Century, 1940.

Hirsch, Jerry. "Behavior Genetics and Individuality Understood." *Science* 142 (1963): 1436–42.

———, ed. *Behavior-Genetic Analysis.* New York: McGraw-Hill, 1967.

———. "Behavior-Genetic Analysis and Its Biosocial Consequences." *Seminars in Psychiatry* 2 (1970): 89–105.

Hobhouse, Leonard. *Social Evolution and Political Theory.* New York: Columbia University Press, 1911.

Hobson, J. A. "Race Eugenics as a Policy." In *Free Thought in the Social Sciences.* New York: Macmillan, 1926.

Hogben, Lancelot. *Genetic Principles in Medicine and Social Science.* London: Williams and Norgate, 1931.

Holmes, Samuel J. *Human Genetics and Its Social Import.* New York: McGraw Hill, 1936.

———. *Studies in Evolution and Eugenics.* New York: Harcourt, Brace, 1923.

Holstein, Jean. *The First Fifty Years at the Jackson Laboratory.* Bar Harbor: The Jackson Laboratory, 1979.

Holtzman, Neil A. *Proceed with Caution: Predicting Genetic Risks in the Recombinant DNA Era.* Baltimore: Johns Hopkins University Press, 1989.

Holtzman, Neil A., and Mark A. Rothstein. "Eugenics and Genetic Discrimination." *American Journal of Human Genetics* 50 (March 1992): 457–59.

Honeyman, Merton, and Ira Gabrielson. "Public Health Aspects of Genetic Screening." *Human Genetics* (Birth Defects Original Articles Series IV) 6 (November 1968): 99–104.

Hubbard, Ruth, and Elijah Wald. *Exploding the Gene Myth.* Boston: Beacon Press, 1993.

Huxley, Julian S. "Eugenics and Society." *Eugenics Review* 28 (1936): 11–31.

———. "Marxist Eugenics." *Eugenics Review* 1 (1936): 66–68.

———. *Memories I.* London: Allen and Unwin, 1907.

Jencks, Christoper. *Inequality: A Reassessment of the Effect of Family and Schooling in America.* New York: Basic Books, 1972.

Jenkins, John B. *Human Genetics.* Menlo Park, Calif.: Benjamin Cummings, 1983.

Jennings, H. S. *The Biological Basis of Human Nature.* New York: W. W. Norton, 1930.

———. *Genetics.* New York: W. W. Norton and Co., 1935.

———. "Health Progress and Race Progress: Are They Incompatible?" *Journal of Heredity* 18 (1927): 271–76.

Jensen, Arthur. "How Much Can We Boost IQ and Scholastic Achievement?" *Harvard Educational Review* 39 (1969): 1–123.

Johnson, Susan. "Eugenics in the Aftermath of WWII: A Study of Articles in *Science, Nature,* and the *Scientific Monthly.*" Unpublished paper, History of Eugenics seminar, University of Massachusetts at Boston, 1995.

Jonas, Gerald. *The Circuit Riders: Rockefeller Money and the Rise of Modern Science.* New York: W. W. Norton Co., 1989.

Joynson, Robert. *The Burt Affair.* London: Routledge, 1989.

Kaback, Michael M. "Heterozygote Screening for the Control of Recessive Disease." In A. Milunsky, ed. *The Prevention of Genetic Disease and Mental Retardation.* Philadelphia: W. B. Saunders, 1975.

Kaempffert, Waldemar. [Review of H. J. Muller, *Out of the Night.*] *New York Times,* March 15, 1936, p. 4.

Kagan, J. S. "Inadequate Evidence and Illogical Conclusion." *Harvard Educational Review* 39 (1969): 274–77.

Kallmann, Franz. "Types of Advice Given by Heredity Counselors." *Eugenics Quarterly* 5 (1958): 48–50.

Kamin, Leon. *The Science and Politics of I.Q.* Potomac, Md.: Lawrence Erlbaum, 1974.

Kammerer, Paul. *The Inheritance of Acquired Characteristics.* Trans. A. Paul Maerker-Branden. New York: Boni and Liveright, 1924. Part B: *Eugenical Part.*

Karjala, Dennis J. "A Legal Research Agenda for the Human Genome Initiative." *Jurimetrics* 32 (Winter, 1992): 121–222.

Kemp, Tage. "Genetic Hygiene and Genetic Counseling." *Acta Genetica et Statistica Medica* 4 (1953): 240–47.

Kennedy, David M. *Birth Control in America: The Career of Margaret Sanger.* New Haven, 1970.

Kevles, Daniel J. *In the Name of Eugenics: Genetics and the Uses of Human Heredity.* New York: Knopf, 1985.

Kimmelman, Barbara A. "An Effort in Reductionist Sociobiology: The Rockefeller Foundation and Physiological Genetics, 1930–1942." Unpublished manuscript, 1981.

Kirkman, N. H. "Projections of a Rebound in Frequency of Mental Retardation from Phenylketonuria." *Applied Research in Mental Retardation* 3 (1982): 319–28.

Kitcher, Philip. *The Lives to Come: The Genetic Revolution and Human Possibilites.* New York: Simon and Schuster, 1996.

——. *Vaulting Ambition: Sociobiology and the Quest for Human Nature.* Cambridge, Mass.: MIT Press, 1985.

Kohler, Robert E. "The Management of Science: The Experience of Warren Weaver and the Rockefeller Foundation Programme in Molecular Biology." *Minerva* 14 (1976): 279–306.

——. "A Policy for the Advancement of Science: The Rockefeller Foundation, 1924–1929." *Minerva* 16 (1978): 480–513.

Koshland, Daniel. "Nature, Nurture, and Behavior." *Science* 235 (March 20, 1987): 1445.

——. Reply to Salvador Luria, *Science,* November 17, 1989, p. 270.

——. "Sequences and Consequences of the Human Genome." *Science,* October 13, 1989, p. 189.

Kraut, Alan M. *Silent Travelers: Germs, Genes, and the Immigrant Menace.* New York: Basic Books, 1994.

Lappe, Mark. *Genetic Politics.* New York: Simon and Schuster, 1979.

Laski, H. J. "The Scope of Eugenics." *Westminster Review* 174 (1910): 34.

Latour, Bruno. "The Impact of Science Studies on Political Philosophy." *Science, Technology, and Human Values* 16 (1991): 3–19.

Latour, Bruno. *We Have Never Been Modern.* Trans. Catherine Porter. Cambridge: Harvard University Press, 1992.

Laudan, Larry. "The Demise of the Demarcation Problem." In Michael Ruse, ed. *But Is It Science?* New York: Prometheus Books, 1988. Pp. 337–50.

Laughlin, H. H. "Report of the Committee to Study and to Report on the Best Practical Means of Cutting off the Defective Germ-Plasm in the American Population. I. The Scope of the Committee's Work." Eugenics Record Office, Bulletin No. 10A. Cold Spring Harbor, N.Y., 1914.

Ledley, F. D. "Differentiating Genetics and Eugenics on the Basis of Fairness." Poster 1818, Eighth International Congress of Human Genetics, Washington, D.C., October 6–11, 1991.

Lehman, Betsy A. "State Will Aid Those with Enzyme Deficiency." *Boston Globe*, January 12, 1994, p. 16.

Lenz, Fritz. "Die Erblichkeit der Geistigen Eigenschaften" In E. Baur *et al. Menschliche Erblichkeitslehre und Rassenhygiene*. Band I: *Menschliche Erblehre*. Munich: J. F. Lehmann, 1936.

Leonard, C. O., *et al*. "Genetic Counseling: A Consumer's View." *New England Journal of Medicine* 287 (1972): 433–39.

Levy, Harvey L. "Nutritional Therapy in Inborn Errors of Metabolism." In Robert J. Desnick. *Treatment of Genetic Diseases*. New York: Churchill Livingstone, 1991. Pp. 1–22.

Lewontin, Richard. "Genetic Aspects of Intelligence." *Annual Review of Genetics* 9 (1975): 387–405.

———. *The Genetic Basis of Evolutionary Change*. New York: Columbia University Press, 1974.

———. "Race and Intelligence." *Bulletin of the Atomic Scientists* 26 (1970): 2–8.

Lippman, Abby. "Prenatal Genetic Testing and Screening: Constructing Needs and Reinforcing Inequalities." *American Journal of Law and Medicine* 17 (1991): 15–50.

Lockwood, Michael. "The Improvement Movement." *Nature,* October 10, 1985, pp. 481–82.

Ludmerer, Kenneth. *Genetics and American Society.* 1973.

Lynch, William T. "Ideology and the Sociology of Scientific Knowledge." *Social Studies of Science* 24 (1994): 197–227.

Mabry, C. C. "Phenylketonuria: Contemporary Screening and Diagnosis." *Annals of Clinical and Laboratory Science* 6 (November–December 1990): 393–97.

MacBride, E. W. "British Eugenists and Birth Control." *Birth Control Review* 6 (1922): 247.

MacFarquhar, Larissa. "Take Back the Nitrous." *Lingua Franca,* May–June 1994, pp. 6–7.

MacKenzie, Donald. "Eugenics in Britain." *Social Studies of Science* 6 (1976): 449–532.

———. "Karl Pearson and the Professional Middle Classes." *Annals of Science* 36 (1979): 125–36.

Macklin, Madge T. "The Value of Medical Genetics to the Clinician." In C. B. Davenport *et al. Medical Genetics and Eugenics.* Philadelphia: Women's Medical College of Pennsylvania, 1940.

Macrakis, Kristie. "The Rockefeller Foundation and German Physics under National Socialism." *Minerva* 27 (1989): 33–57.

Manicol, A. M., *et al.* "Implications of a Genetic Screening Programme for Polycystic Kidney Disease." *Aspects of Renal Care* 1 (1986): 219–22.

Manuel, Diane. "Failing Grades for Textbooks." *Boston Globe,* March 1, 1992, pp. A1, 9–10.

Marfatia, Lavanya, Diana Punales-Morejon, and Rayna Rapp. "Counseling the Underserved: When an Old Reproductive Technology Becomes a New Reproductive Technology." *Birth Defects: Original Article Series* 26 (1990): 109–26.

Masland, Richard L. "The Prevention of Mental Subnormality." In R. L. Masland *et al., Mental Subnormality: Biological, Psychological, and Cultural Factors.* New York: Basic Books, 1958.

McDonough, Terrence, and Joseph Eisenhauer. "Sir Robert Giffen and the Great Potato Famine: A Discussion of the Role of a Legend in Neoclassical Economics." *Journal of Economic Issues* 29 (September 1995): 747–59.

McInerney, Joseph D. "Why Biological Literacy Matters: A Review of Commentaries Related to *The Bell Curve: Intelligence and Class Structure in American Life," Quarterly Review of Books* 71 (1996): 81–96.

McKenzie, Richard B. "The Emergence of Kickbacks in University Textbook Adoptions." Unpublished manuscript, 1987.

McKusick, V. A. *Mendalian Inheritance in Man: Catalogue of Autosomal Dominant, Autosomal Recessive, and X-Linted Phenotypes.* 10th ed. Baltimore: Johns Hopkins University Press, 1992.

Meads, Marston. *The Miracle at Hawthorne Hill.* Winston-Salem: Medical Center of Bowman-Gray School of Medicine and North Carolina Baptist Hospital, 1988.

Medawar, P. B., and J. S. Medawar. *The Life Sciences: Current Ideas in Biology.* New York: Harper and Row, 1977.

Mehler, Barry. "The New Eugenics: Academic Racism in the U.S. Today." *Science for the People* 15 (May–June 1983): 18–23.

Mill, John Stuart. "On Liberty." *Essential Works of John Stuart Mill.* New York: Bantam, 1855; rpt. 1961.

———. *Principles of Political Economy.* New York: 1848, Augustus M. Kelley, 1969; reprint of 1909 edition.

Miller, Adam. "Professors of Hate." In R. Jacoby and N. Glauberman, eds. *The Bell Curve Debate: History, Documents, Opinions.* New York: Times Books, 1994. Pp. 162–78.

Millner, Bernard N. "Insurance Coverage of Special Foods Needed in the Treatment of Phenylketonuria." *Public Health Reports* 108 (January–February 1993): 60–65.

"Modified Proposal for a Council Decision Adopting a Specific Research and Technological Development in the Field of Health: Human Genome Analysis (1990–1991)." *Official Journal of the European Communities,* CB-CO-89-485-EN-C, Brussels, November 13, 1989.

Montagu, Ashley. *Human Heredity.* Cleveland: World Publishing, 1959.

Mooney, Gavin, and Mette Lange. "Ante-natal Screening: What Constitutes 'Benefit'?" *Social Science and Medicine* 37 (1993): 873.

Morgan, T. H. *Evolution and Genetics.* 2d ed. Princeton: Princeton University Press, 1925.

Morris, Charles G. *Psychology: An Introduction.* Englewood Cliffs: Prentice-Hall, 1985.

Müller-Hill, Benno. *Murderous Science.* Trans. George Fraser. New York: Oxford University Press, 1988.

Muller, H. J. "The Dominance of Economics over Eugenics." *Birth Control Review* 16 (1932): 236–38.

———. "Our Load of Mutations." *American Journal of Human Genetics* 2 (1950): 111–76.

———. *Out of the Night.* New York: Vanguard Press, 1935.

Murphy, Gardner. "A Research Program for Qualitative Eugenics." *Eugenics Quarterly* 1 (1954): 209–12.

Natowicz, Marvin R., and Joseph S. Alper. "Genetic Screening: Triumphs, Problems, and Controversies." *Journal of Public Health Policy* 12 (1991): 475–91.

Neel, James V. "Lessons from a Primitive People." *Science* 170: 820–21.

———. *Physician to the Gene Pool.* New York: Wiley, 1994.

———. "On Emphases in Human Genetics." *Genetics* 78 (1974): 35–40.

Neel, James V., and William J. Schull. *Human Heredity.* Chicago: University of Chicago Press, 1954.

Nelkin, Dorothy, and Laurence Tancredi. *Dangerous Diagnostics: The Social Power of Biological Information.* New York: Basic Books, 1989.

"Notes of the Quarter." *Eugenics Review* 27 (1935): 188.

Nozick, Robert. *Anarchy, State, and Utopia.* New York: Basic Books, 1974.

Oliver, Clarence P. "Human Genetics Program at the University of Texas." *Eugenical News* 37 (1952): 25–31.

———. "A Report on the Organization and Aims of the Dight Institute." Dight Bulletin #1 (1943), p. 2.

———. "Statement." In "A Panel Discussion: Genetic Counseling." *American Journal of Human Genetics* 4 (1953): 343.

Oliver, Kelly. "Keller's Gender/Science System: Is the Philosophy of Science to Science as Science Is to Nature?" *Hypatia* 3 (1989): 137–48.

[Osborn, Frederick]."Editorial." *Eugenics Quarterly* 1 (1954): 2.

Osborn, Frederick. *The Future of Human Heredity.* New York: Harper and Brothers, 1968.

———. "Heredity and Practical Eugenics Today." *Eugenical News* 33 (1948): 1–6.

———. *Preface to Eugenics.* 2d ed. New York: Harper and Brothers, 1951.

———. Transcript, Oral History Interview, July 10, 1974, Columbia University, New York.

Pai, Anna, and Helen Marcus-Roberts. *Genetics: Its Concepts and Implications.* Englewood Cliffs, N.J.: Prentice-Hall, 1981.

Papalia, D. E., and S. W. Olds. *Psychology.* New York: McGraw-Hill, 1985.

Pastore, Nicholas. *The Nature-Nurture Controversy.* New York: Kings Crown Press, 1949.

Paul, Diane B. *Controlling Human Heredity: 1865 to the Present.* Atlantic Highlands, N.J.: Humanities Press, 1995.

———. "Dobzhansky in the 'Nature-Nurture' Debate." In M. Adams, ed. *The Evolution of Theodosius Dobzhansky.* Princeton: Princeton University Press, 1994. Pp. 219–31.

———. "'In the Interests of Civilization': Marxist Views of Race and Culture in the Nineteenth Century." *Journal of the History of Ideas* 32 (January 1981): 115–38.

———. "Is Human Genetics Disguised Eugenics?" In Robert F. Weir *et al.,* eds. *Genes and Human Self-Knowledge: Historical and Philosophical Reflections on Modern Genetics.* Iowa City: University of Iowa Press, 1994. Pp. 67–83.

———. "The Market as Censor." *PS: Political Science and Politics* 21 (1988): 31–35.

———. "Marxism, Darwinism, and the Theory of Two Sciences." *Marxist Perspectives* 2 (Spring 1979): 116–43.

———. "The Nine Lives of Discredited Data," *The Sciences* 27 (1987): 26–30.

———. "'Our Load of Mutations' Revisited." *Journal of the History of Biology* 20 (1987): 321–35.

———. "The Rockefeller Foundation and the Origins of Behavior Genetics." In K. Benson *et al.,* eds. *The Expansion of American Biology.* New Brunswick, N.J.: Rutgers University Press, 1991. Pp. 262–83.

———. "Textbook Treatments of the Genetics of Intelligence." *Quarterly Review of Biology* 60 (1985): 317–26.

———. "A War on Two Fronts: J. B. S. Haldane and the Response to Lysenkoism in Britain." *Journal of the History of Biology* 16 (Spring 1983): 1–37.

Paul, Diane B., and Arthur R. Blumenthal. "On the Trail of Little Albert." *Psychological Record* 39 (1989): 547–53.

Paul, Diane B., and Paul J. Edelson. "The Struggle over Screening." In Soraya de Chadarevian and Harmke Kamminga, eds. *Molecularising Biology and Medicine: New Practices and Alliances, 1930s–1970s.* Reading, U.K.: Harwood Academic, 1998.

Paul, Diane B., and Raphael Falk. "Scientific Responsibility and Political Context: The Case of Genetics under the Swastika." In Jane Maienschein and Michael Ruse, eds. *Biology and the Foundations of Ethics.* Cambridge: Cambridge University Press, in press.

Paul, Diane B., and Barbara A. Kimmelman. "Mendel in America: Theory and Practice, 1900–1919." In R. Rainger *et al.*, eds. *The American Development of Biology.* Philadelphia: University of Pennsylvania Press, 1988.

Paul, Eden. "Eugenics, Birth Control, and Socialism." In Eden Paul and Cedar Paul, eds. *Population and Birth-Control: A Symposium.* New York: Critic and Guide, 1917.

Pauling, Linus. "Reflections on the New Biology: Foreword." *UCLA Law Review* 15 (1968): 267–72.

Pearson, Karl. *The Ethic of Free Thought.* London: Adam and Charles Black, 1901.

———. "Mendelism and the Problem of Mental Defect. III. On the Graduated Character of Mental Defect and on the Need for Standardizing Judgments as to the Grade of Social Inefficiency which Shall Involve Segregation." In *Questions of the Day and Fray.* No. 9. London: Dulau, 1914.

———. "The Moral Basis of Socialism." In *The Ethic of Free Thought.* London, 1887; rpt. 1901.

———. *The Problem of Practical Eugenics.* London: Dulau, 1912.

Pencarinha, Deborah, *et al.* "A Study of the Attitudes and Reasoning of M.S. Genetic Counselors Regarding Ethical Issues in Medical Genetics." Poster Presentation #1821, 12th International Congress of Human Genetics, Washington, D.C., October 6–11, 1991.

Penfield, Wilder. *The Difficult Art of Giving: The Epic of Alan Gregg.* Boston: Little, Brown, 1967.

Penrose, Lionel S. *Biology of Mental Defect.* London: Sidgwick and Jackson, 1949.

———. "Phenylketonuria: A Problem in Eugenics." *The Lancet,* June 29, 1946, pp. 949–53.

Pernick, Martin S. *The Black Stork: Eugenics and the Death of 'Defective' Babies in American Medicine and Motion Pictures.* New York and Oxford: Oxford University Press, 1996.

Phelps, G. "The Eugenics Crusade of Charles Fremont Dight." *Minnesota History* 49 (1984): 99–108.

Pick, Daniel. *Faces of Degeneration: A European Disorder, c.1848–c.1918.* Cambridge: Cambridge University Press, 1989.

Pius XI. "On Christian Marriage." The English Translation. New York: Barry Vail, 1931.

Plomin, Robert. "The Role of Inheritance in Behavior." *Science* 248 (1990): 183–84.

Pollitt, Katha. "Pomolotov Cocktail." *The Nation,* June 10, 1996, p. 9.

Popenoe, P. "Feeblemindedness." *Journal of Heredity* 6 (1915): 32–36.

Popenoe, P., and R. H. Johnson. *Applied Eugenics.* New York: Macmillan, 1918.

Porter, Ian. "Evolution of Genetic Counseling in America." In H. A. Lubs and F. de la Cruz, eds. *Genetic Counseling.* New York: Raven Press, 1977.

"President's Summary of the Discussion." *British Medical Journal,* April 11, 1923, pp. 233–34.

Press, Nancy A., and Carol H. Browner. "Collective Silences, Collective Fictions: How Prenatal Testing Became Part of Routine Prenatal Care." In K. H. Rothenberg and E. J. Thomson. *Women and Prenatal Testing: Facing the Challenge of Genetic Technology.* Columbus: Ohio State University Press, 1994. Pp. 201–18.

"Price List and Order Form." Dietary Specialties, Inc.

Proctor, Robert. *Racial Hygiene: Medicine under the Nazis.* Cambridge, Mass.: Harvard University Press, 1988.

———. *Value-Free Science? Purity and Power in Modern Knowledge.* Cambridge: Harvard University Press, 1991.

"Progressive Parties and Eugenics." *Eugenics Review* 28 (1936): 296.

Provine, W. B. *The Origins of Theoretical Population Genetics.* Chicago: University of Chicago Press, 1971.

Prytula, Robert E., Gerald D. Oster, and Stephen F. Davis. "The 'Rat-Rabbit' Problem: What Did John B. Watson Really Do?" *Teaching of Psychology* 4 (1977): 44–46.

Punnett, R. C. "As a Biologist Sees It." *The Nineteenth Century* 97 (1925): 697–707.

———. "Eliminating Feeblemindedness." *Journal of Heredity* 8 (1917): 464–65.

———. "Genetics and Eugenics." *Problems in Eugenics: Papers Communicated to the First International Eugenics Congress.* London: Eugenics Education Society, 1912.

———. *Mimicry in Butterflies.* Cambridge: Cambridge University Press, 1915.

Rapp, Rayna. "Chromosomes and Communication: The Discourse of Genetic Counseling," *Medical Anthropology Quarterly* 2 (1988): 152.

Rathus, S. A. *Psychology.* New York: Holt, Rinehart, and Winston, 1987.

Reed, Julia. "Genes: Little Things That Mean a Lot." *U.S. News and World Report,* December 15, 1986, p. 8.

Reed, Sheldon. *Counseling in Medical Genetics.* Philadelphia: W. B. Saunders, 1955.

———. "Heredity Counseling." *Eugenics Quarterly* 1 (1954): 48–49.

———. "Heredity Counseling and Research." *Eugenical News* 37 (1952): 43.

———. *Parenthood and Heredity* (New York: John Wiley and Sons, 1964.

———. "A Short History of Genetic Counseling." *Dight Institute Bulletin* #14 (1974), pp. 4–5.

Reilly, Philip R. *The Surgical Solution: A History of Involuntary Sterilization in the United States.* Baltimore: Johns Hopkins University Press, 1991.

Robbins, Bruce, and Andrew Ross. "Mystery Science Theater." *Lingua Franca,* July–August 1996, pp. 54–57.

Roberts, Elmer. "Biology and Social Problems," *Dight Institute Bulletin* #4 (1946): 18.

Robertson, John. "Embryos, Families, and Procreative Liberty: The Legal Structures of the New Reproduction." *Southern California Law Review* 59 (1986): 942–1041.

———. "Procreative Liberty and the Control of Contraception, Pregnancy and Childbirth." *Virginia Law Review,* 69 (1983): 405–62.

Rockefeller Foundation. "President's Review." 1936.

Rogers, Carl R. *Client-Centered Therapy: Its Current Practice, Implications, and Theory.* Boston: Houghton Mifflin, 1951.

Rose, Steven. "Environmental Effects on Brain and Behaviour." In Ken Richardson *et al. Race, Culture and Intelligence.* Baltimore: Penguin Books, 1972.

Rosenberg, Charles. Review of Daniel J. Kevles, *In the Name of Eugenics. Journal of American History,* 73 (June 1986): 232–33.

Russell, Bertrand. "Eugenics." In *Marriage and Morals.* London: G. Allen and Unwin, 1924. Pp. 255–73.

Ryan, Maura A. "The Argument for Unlimited Procreative Liberty: A Feminist Critique." *Hastings Center Report,* July–August 1990, pp. 6–12.

Saletan, William. "Genes 'R Us." *New Republic,* July 16 and 24, 1989, pp. 18–20.

Samelson, Franz. "J. B. Watson's Little Albert, Cyril Burt's Twins, and the Need for a Critical Science." *American Psychologist* 35 (1980): 619–25.

———. "Organizing the Kingdom of Behavior." *Journal of the History of the Behavioral Sciences* 21 (1985): 33–47.

Schuett, Virginia E. *National Survey of Treatment Programs for PKU and Selected Other Inherited Metabolic Disorders.* Rockville, Md.: Bureau of Maternal and Child Health and Resources Development. Public Health Services. U. S. Dept. of Health and Human Services, 1990.

Schuett, Virginia E., et al. "Diet Discontinuation Policies and Practices of PKU Clinics in the United States." American Journal of Public Health 70 (1980):498.

Scott, John Paul. "Investigative Behavior: Toward a Science of Sociality." In D. A. Dewsbury, ed. Leaders in the Study of Animal Behavior: Autobiographical Perspectives. Lewisburg, Penn.: Bucknell University Press, 1985.

Scott, John Paul, and John L. Fuller. "Heredity and the Social Behavior of Mammals." The Roscoe B. Jackson Memorial Laboratory, 27th Annual Report, 1955–1956.

————. Genetics and the Social Behavior of the Dog. Chicago: University of Chicago Press, 1965.

Scriver, Charles R. "Phenylketonuria—Genotypes and Phenotypes." New England Journal of Medicine, May 2, 1991, pp. 1280–81.

Searle, G. R. "Eugenics and Class." In Charles Webster, ed. Biology, Medicine and Society 1840–1940. Cambridge: Cambridge University Press, 1981. Pp. 217–42.

Sedgwick, J. "Inside the Pioneer Fund." In R. Jacoby and N. Glauberman, eds. The Bell Curve Debate: History, Documents, Opinions. New York: Times Books, 1994. Pp. 144–61.

Seligman, Daniel. A Question of Intelligence: The IQ Debate in America. New York: Birch Lane Press, 1992.

Shapin, Steven. "Here and Everywhere: Sociology of Scientific Knowledge." Annual Review of Sociology 21 (1995): 289–321.

Shaw, George Bernard. Sociological Papers. London: Macmillan 1905.

Shaw, Margery W. "Conditional Prospective Rights of the Fetus." Journal of Legal Medicine 5 (1984): 63–116.

————. "The Potential Plaintiff: Preconception and Prenatal Torts." In A. Milunsky and G. Annas, eds., Genetics and the Law II. New York: Plenum Press, 1980.

————. "To Be or Not to Be? That Is the Question," American Journal of Human Genetics 36 (1984): 1–9.

Singer, Eleanor. "Public Attitudes toward Genetic Testing." Population Research and Policy Review 10 (1991): 235–55.

Smith, Charles. "Ascertaining Those at Risk in the Prevention and Treatment of Genetic Disease." In Alan E. H. Emery, ed. Modern Trends in Human Genetics I. New York: Appleton-Century-Crofts, 1972. Pp. 350–69.

Smith, I., et al. "Effect of Stopping Low-Phenylalanine Diet on Intellectual Progress of Children with Phenylketonuria." British Medical Journal 2 (1978): 723–26.

Snow, C. P. "Revolution in Ourselves." Spectator 157 (1936): 64.

Snyder, Lawrence. "Heredity and Modern Life." In R. G. Gates et al., eds. Medical Genetics and Eugenics. Vol. 2. Philadelphia: Women's Medical College of Pennsylvania, 1943.

Snyderman, Mark, and Stanley Rothman. *The IQ Controversy.* New Brunswick, N.J.: Transaction Books, 1988.

"Social Biology and Population Improvement." *Nature* 144 (1939): 521.

Sokal, Alan. "A Physicist Experiments with Cultural Studies." *Lingua Franca*, May–June 1996, pp. 62–64.

Sokal, David C., *et al.* "Prenatal Chromosomal Diagnosis: Racial and Geographic Variation for Older Women in Georgia." *Journal of the American Medical Association* 244 (1980): 1355–57.

Soloway, Richard A. *Birth Control and the Population Question in England, 1877–1930* (Chapel Hill: University of North Carolina Press, 1982.

———. *Demography and Degeneration: Eugenics and the Declining Birthrate in Twentieth-Century Britain.* Chapel Hill: University of North Carolina Press, 1990.

Sommer, Robert, Marina Estabrook, and Karen Horobin. "Faculty Awareness of Textbook Prices." *Teaching of Psychology* 15 (1988): 17–21.

Sorenson, James R. "Genetic Counseling: Values that Have Mattered." In D. M. Bartels *et al.*, eds. *Prescribing Our Future: Ethical Challenges in Genetic Counseling.* New York: Aldine de Gruyter, 1993. Pp. 3–14.

Sorenson, J. R., and A. J. Culbert. "Genetic Counselors and Counseling Orientation—Unexamined Topics in Evaluation." In H. A. Lubs and F. de la Cruz, eds. *Genetic Counseling.* New York: Raven Press, 1974. Pp. 131–54.

Stern, Curt. *Principles of Human Genetics.* San Francisco: Freeman, 1949.

———. *Principles of Human Genetics.* New York: Freeman, 1949.

Stocks, J. L. [Review of H. J. Muller, *Out of the Night.*] *Manchester Guardian,* June 9, 1936, p. 7;

Sturtevant, A. H. "Social Implications of the Genetics of Man." *Science* 120 (1954): 405–407.

Sujansky, Eva, *et al.* "Attitudes of At-Risk and Affected Individuals Regarding Presymptomatic Testing for Autosomal Dominant Polycystic Kidney Disease." *American Journal of Medical Genetics* 35 (1990): 510–15.

Sutton, H. Eldon. *Genes, Enzymes, and Inherited Diseases.* New York: Holt, Rinehart, and Winston, 1961.

———. *Introduction to Human Genetics.* San Francisco: W. H. Freeman, 1980.

Suzuki, David, and Peter Knudtson. *Genethics: The Clash Between the New Genetics and Human Values.* Cambridge, Mass.: Harvard University Press, 1989.

Swanson, E. "Biographical Sketch of Charles Fremont Dight." *Dight Institute Bulletin* #1 (1943), pp. 9–22.

Swazey, Judith P. "Phenylketonuria: A Case Study in Biomedical Legislation." *Journal of Urban Law* 48 (1971): 883–931.

Tamarin, Robert H. *Principles of Genetics.* Boston: Willard Grant, 1982.

Taylor, Charles. *Philosophical Arguments.* Cambridge: Harvard University Press, 1995.

Taylor, Howard F. *The IQ Game: A Methodological Inquiry into the Heredity-Environment Controversy.* New Brunswick, N.J.: Rutgers University Press, 1980.

Twiss, Seymour. "The Genetic Counselor as Moral Advisor." *Birth Defects Original Articles Series 15* (1979).

U.S. Congress, Office of Technology Assessment. *Cystic Fibrosis and DNA Tests: Implications of Carrier Screening.* OTA-BA-532. Washington, D.C.: U.S. Government Printing Office, August 1992.

———. *Mapping Our Genes.* Washington, D.C.: GPO, 1988.

Waisbren, Susan, et al. "The New England Maternal PKU Project: Identification of At-Risk Women." *American Journal of Public Health 78* (July 1988): 789–92.

Walker, Herbert Eugene. *Genetics: An Introduction to the Study of Heredity.* New York: Macmillan, 1913.

Wallas, Graham. *The Great Society: A Psychological Analysis.* New York: Macmillan, 1914.

Ward, Harold. [Review of H. J. Muller, *Out of the Night.*] *New Republic 86* (1936): 284.

Watson, James D. "The Human Genome Project: Past, Present, and Future." *Science,* April 6, 1990, pp. 244–48.

Watson, John B., and Rosalie Rayner. "Conditioned Emotional Reactions." *Journal of Experimental Psychology 3* (1920): 1–14.

Watson, John B., and Rosalie Rayner Watson, "Studies in Infant Psychology." *Scientific Monthly 13* (1921): 493–514.

Watson, M. L., et al. "Adult Polycystic Kidney Disease." *British Medical Journal 300* (1990): 62–63.

Webb, Sidney. *The Decline in the Birth-Rate.* London: Fabian Society, 1907.

———. "Eugenics and the Poor Law: The Minority Report. *Eugenics Review 2* (1910–1911): 233–41.

———. *Genetic Principles in Medicine and Social Science.* London: Williams and Norgate, 1931.

Webb, Sidney, and Beatrice Webb. *The Prevention of Destitution.* London: Longmans, Green, 1911.

Weglage, J., et al. "Psychological and Social Findings in Adolescents with Phenylketonuria." *European Journal of Pediatrics 151* (July 1992): 522–25.

Weindling, Paul. "The Rockefeller Foundation and German Biomedical Sciences, 1920–40: From Educational Philanthropy to International Science Policy." In Nicolaas A. Rupke, ed. *Science, Politics and the Public Good: Essays in Honour of Margaret Gowing.* London: Macmillan Press, 1988. Pp. 119–40.

Wells, H. G. *A Modern Utopia.* New York: Charles Scribner's Sons, 1905.

———. *Sociological Papers.* London: Macmillan, 1905.

Werskey, Gary. *The Visible College: A Collective Biography of British Scientific Socialists of the 1930s.* New York: Holt, Rinehart, and Winston, 1978.

Wertz, Dorothy, et al. "Attitudes toward Abortion among Parents of Children with Cystic Fibrosis." American Journal of Public Health 81 (1991): 992–96.

Wertz, Dorothy, and John Fletcher. "Ethical Decision Making in Medical Genetics: Women as Patients and Practitioners in Eighteen Nations." In K. Ratcliff et al. Healing Technology: Feminist Perspectives. Ann Arbor: University of Michigan Press, 1989.

Wertz, Dorothy C., and John C. Fletcher, eds. Ethics and Human Genetics: A Cross-Cultural Perspective. New York: Springer-Verlag, 1989.

———. "Fatal Knowledge? Prenatal Diagnosis and Sex Selection." Hastings Center Report, May–June 1989, pp. 21–27.

Wertz, Dorothy C., John C. Fletcher, and John J. Mulvihill. "Medical Geneticists Confront Ethical Dilemmas: Cross-cultural Comparisons among 18 Nations." American Journal of Human Genetics 46 (1990): 1200–13.

Whitney, P. W. "Communist Eugenics." Journal of Heredity 27 (1936): 132–35.

Wiggam, Albert E. Fruit of the Family Tree. Indianapolis: Bobbs-Merrill, 1924.

Williams, Juan. "Violence, Genes, and Prejudice." Discover, November 1994, pp. 93–102.

Wingerson, Lois. Mapping our Genes. New York: Dutton, 1990.

Winkler, Karen J. "New Approaches Changing the Face of Textbook Publishing." Chronicle of Higher Education, May 16, 1977, pp. 1, 10.

Wolpert, Louis. The Unnatural Nature of Science: Why Science Does Not Make (Common) Sense. London: Faber and Faber, 1992.

Woolgar, Steve, and Keith Grint. "A Further Decisive Refutation of the Assumption That Political Action Depends on the 'Truth' and a Suggestion That We Need to Go beyond This Level of Debate: A Reply to Rosalind Gill." Science, Technology, and Human Values 21 (1996): 353–57.

Wright, Lawrence. "Double Mystery." The New Yorker, August 7, 1995, pp. 44–62.

Wright, Robert. "Achilles' Helix." New Republic, July 9 and 16, 1990, pp. 21–31.

———. "Dumb Bell." New Republic, January 2, 1995, p. 6.

Yerkes, Robert M., ed. Psychological Examining in the United States Army. Vol. 15. Memoirs of the National Academy of Sciences. Washington, D.C., 1921.

Zenderland, Leila. "On Interpreting Photographs, Faces, and the Past." American Psychologist 43 (September 1988): 743–44.

———. "A Sermon of New Science: The Kallikak Family as Eugenic Parable." Paper presented at the History of Science Society Annual Meeting, Washington, D.C., 1994.

INDEX

Jervis, George, 173–74
Johns Hopkins University, 48, 67
Johnson, R. H., 119, 129
Jonas, Gerald, 56
Journal of the American Medical Association (JAMA), 23

Kaback, Michael, 166
Kaempffert, Walter, 72
Kaiser Wilhelm Institute (KWI) for Anthropology, Human Genetics, and Eugenics, 61
KWI for Brain Research, 61
Kallikak, Deborah, 158
The Kallikak Family. See "Goddard, Henry H."
Kallikak, Martin, Sr., 159–60, 167
Kallmann, Franz, 63, 137, 138, 139, 162–63
Kamin, Leon, 38, 47, 51n, 86, 89
The Science and Politics of I.Q., 84, 85
Kammerer, Paul, 13
Karjala, Dennis, 102
Keeler, Clyde, 72
Kemp, Tage, 60, 137
Kevles, Daniel J., 11, 30n, 32n, 103
In the Name of Eugenics, 103
Khayam, Omar, 19
Kitcher, Philip, 96
The Lives to Come, 96
Kohler, Robert, 54
Koshland, Daniel J., 82, 90
Kraepelin, Emil, 55
KWI. *See* "Kaiser Wilhelm Institute"

laissez-faire, 14
Lal, Gobind, 72
Lamarckism, 20, 23
Lambroso, 159
Laski, H. J., 13, 30n
Latour, Bruno, 7
Laughlin, Harry, 62
Laura Spelman Rockefeller Memorial, 75n

Law for the Prevention of Genetically Diseased Offspring, 144, 162–63
Lebensborn project, 100
See also "Hitler, Adolf"
Left Book Club, 17
Lejune, Jerome, 141
Lenin, Vladimir, 19–21
Lenz, Fritz, 144
Leonard, Claire, 147
Leonardo da Vinci, 19
Levan, Albert, 141
Levit, Solomon, 24
Lewontin, Richard C., 27, 35n, 84
Lillie, Frank, 57, 58
Lippman, Abby, 100
"Little Albert," 37, 47–49, 50n
Little, Clarence C., 53, 57, 67–73
Locke, John, 85
Lockwood, Michael, 98
Ludmerer, Kenneth, 12
Lysenko, T. D., 28
Lysenko controversy, 13, 25
Lysenkoism, 11, 25

MacKenzie, Donald, 15
Macklin, Madge, 135, 137, 138, 145
manic depression, 144
Marcus-Robert, Helen, 47, 51n
Marks, Joan, 154n
Marx, Karl, 16, 19
Marxism, 2, 3, 8–9, 11, 12, 13, 15–17, 25, 28, 30n, 33n, 105
Masland, Richard, 140
mating: assortive, 125
random, 123, 124, 125
Mason, Max, 56, 57, 58
McConnell, Campbell R., 44
McGraw-Hill, 51n
McGregor Foundation, 74, 137
Medawar, J. S., 118
Medawar, P. B., 118
Medicaid, 179–80
medical genetics, 64, 65, 95, 100, 102–103, 106, 133–37, 140–41, 147, 153n, 164, 167